W9-CMB-956

PRO/ENGINEER 2001 ASSISTANT

PRO/ENGINEER 2001 ASSISTANT

David S. Kelley
Purdue University

Boston   Burr Ridge, IL   Dubuque, IA   Madison, WI   New York   San Francisco   St. Louis
Bangkok   Bogotá   Caracas   Kuala Lumpur   Lisbon   London   Madrid   Mexico City
Milan   Montreal   New Delhi   Santiago   Seoul   Singapore   Sydney   Taipei   Toronto

McGraw-Hill Higher Education

*A Division of The **McGraw-Hill** Companies*

PRO/ENGINEER 2001 ASSISTANT

Published by McGraw-Hill, a business unit of The McGraw-Hill Companies, Inc., 1221 Avenue of the Americas, New York, NY 10020. Copyright " 2002 by The McGraw-Hill Companies, Inc. All rights reserved. No part of this publication may be reproduced or distributed in any form or by any means, or stored in a database or retrieval system, without the prior written consent of The McGraw-Hill Companies, Inc., including, but not limited to, in any network or other electronic storage or transmission, or broadcast for distance learning.

Some ancillaries, including electronic and print components, may not be available to customers outside the United States.

This book is printed on recycled, acid-free paper containing 10% postconsumer waste.

1 2 3 4 5 6 7 8 9 0 QPD/QPD 0 9 8 7 6 5 4 3 2 1

ISBN 0–07–249939–7

General manager: *Thomas E. Casson*
Publisher: *Elizabeth A. Jones*
Senior developmental editor: *Kelley Butcher*
Executive marketing manager: *John Wannemacher*
Project manager: *Joyce M. Berendes*
Senior production supervisor: *Sandy Ludovissy*
Coordinator of freelance design: *Rick D. Noel*
Cover illustration: *©Vehicle Research Institute/Dr. Michael Seal/William Connelly*
Cover designer: *Rick D. Noel*
Supplement producer: *Brenda A. Ernzen*
Media technology senior producer: *Phillip Meek*
Compositor: *Interactive Composition Corporation*
Typeface: *10/12 Times Roman*
Printer: *Quebecor World Dubuque, IA*

Library of Congress Cataloging-in-Publication Data

Kelley, David S.
    Pro/Engineer 2001 assistant / David S. Kelley. — 1st ed.
cm. — (The McGraw-Hill graphics series)
    Includes index.
    ISBN 0-07-249939-7
    1. Pro/ENGINEER. 2. Computer-aided design. 3. Mechanical drawing. I. Title.
    II. Series.

TA174 .K444   2002
620′.0042—dc21   2001034102
                                                     CIP

www.mhhe.com

This book is dedicated to my parents, for their love and support.

PREFACE

PURPOSE

My decision to write a Pro/ENGINEER textbook was based on the lack of a comprehensive textbook on this popular computer-aided design package. I focused on several objectives and ideas when I started to develop this project:

1. To write a textbook for an introductory course in engineering graphics.
2. To meet the needs of institutions teaching a course on parametric design and constraint-based modeling.
3. To create a book that would serve as a self-paced, independent study guide for the learning of Pro/ENGINEER for those who do not have the opportunity to take a formal course.
4. To incorporate a tutorial approach to the learning of Pro/ENGINEER in conjunction with detailed reference material.
5. To include topics that make the text a suitable supplement for an upper division course in mechanical design.

APPROACHES TO USING THE TEXTBOOK

This textbook is designed to serve as a tutorial, reference, and lecture guide. Chapters start by covering selected topics in moderate detail. Following the reference portion of each chapter are one or more tutorials covering the chapter's objectives and topics. At the end of each chapter are practice problems used to reinforce concepts covered in the chapter and previously in the book.

I had several ideas in mind when developing this approach to the book. Since Pro/ENGINEER is a menu-intensive, computer-aided design application, the most practical pedagogical method to cover Pro/ENGINEER's capabilities (that would be the most beneficial both to students and instructors) would be to approach this book as a tutorial. In addition to the provided tutorials, this book provides detailed reference material. A typical approach to teaching Pro/ENGINEER would be to provide a tutorial exercise followed by a nontutorial practice or practical problem. Usually students can complete the tutorial, but they may run into problems on the practice exercise. One of the problems that Pro/ENGINEER students have is digging back through the tutorial to find the steps for performing specific modeling tasks. The reference portion of each chapter in this text provides step-by-step guides for performing specific Pro/ENGINEER modeling tasks outside of a tutorial environment.

STUDENTS OF PRO/ENGINEER

One of the objectives of this book is to serve as a stand-alone text for independent learners of Pro/ENGINEER. This book is approached as a tutorial to help meet this objective. Since Pro/ENGINEER is menu intensive, tutorials in this book use numbered steps to guide the selection of menu options. The following is an example of a tutorial step:

STEP 6: **Place Dimensions according to Design Intent.**

Use the Dimension icon to match the dimensioning scheme shown in Figure 4–24. Placement of dimensions on a part should match design intent. With Intent Manager activated (Sketch >> Intent Manager), dimensions and constraints are provided automatically that fully define the section. Pro/ENGINEER does not know what dimensioning scheme will match design intent, though. Due to this, it is usually necessary to change some dimension placements.

MODELING POINT If possible, a good rule of thumb to follow is to avoid modifying the section's dimension values until your dimension placement scheme matches design intent.

The primary menu selection is shown in bold. In this example, you are instructed to use the dimension option (portrayed by the Dimension icon) to create dimensions that match the part's design intent. Following the specific menu selection, when appropriate, is the rationale for the menu selection. In addition, Modeling Points are used throughout the book to highlight specific modeling strategies.

CHAPTERS

The following is a description and rationale for each chapter in the book:

CHAPTER 1 PRO/ENGINEER'S USER INTERFACE

This chapter covers basic principles behind Pro/ENGINEER's interface and menu structure. The purpose is to serve as a guide and reference for later modeling activities. A tutorial is provided to reinforce the chapter's objectives.

CHAPTER 2 EXTRUDING, MODIFYING, AND REDEFINING FEATURES

Chapter 2 is the first chapter covering Pro/ENGINEER's solid modeling capabilities. Pro/ENGINEER's protrusion and cut commands are introduced and the extrude option is covered in detail. In addition, modification and datum plane options are introduced.

CHAPTER 3 FEATURE CONSTRUCTION TOOLS

While the protrusion and cut commands are Pro/ENGINEER's basic tools for creating features, this chapter covers additional feature creation tools. Covered in detail are the hole, round, rib, and chamfer commands; creating draft surfaces; shelling a part; cosmetic features; and creating linear patterns.

CHAPTER 4 REVOLVED FEATURES

Many Pro/ENGINEER features are created by revolving around a center axis. Examples include the revolve option found under the protrusion and cut commands and the sketched hole option. These options, along with creating rotational patterns and datum axes, are covered in this chapter.

CHAPTER 5 FEATURE MANIPULATION TOOLS

Pro/ENGINEER provides tools for manipulating existing features. Manipulation tools covered include the group option, copying features, and creating relations.

CHAPTER 6 CREATING A PRO/ENGINEER DRAWING

Since Pro/ENGINEER is primarily a modeling and design application, the creation of engineering drawings is considered a downstream task. Despite this, there is a need to cover the capabilities of Pro/ENGINEER's drawing mode. This chapter covers the creation of general and projection views.

CHAPTER 7 SECTIONS AND ADVANCED DRAWING VIEWS

Due to the length and depth of Chapter 8, the creation of section and auxiliary views is covered in a separate chapter.

ACKNOWLEDGMENTS

I would like to thank many individuals for contributions provided during the development of this text. I would like to thank faculty, friends, and students at Purdue University and Western Washington University. I am grateful to McGraw-Hill for producing the beta edition of this book.

The editorial and production groups of McGraw-Hill have been wonderful to work with. I would especially like to thank Betsy Jones, Kelley Butcher, Joyce Berendes, and Melinda Dougharty. Thanks also to Gary Bertoline and Craig Miller at Purdue University for their support, help, and encouragement.

TRADEMARK ACKNOWLEDGMENTS

The following are registered trademarks of Parametric Technology Corporation ®: PTC, Pro/ENGINEER, Pro/INTRALINK, Pro/MECHANICA, Windchill, and most other applications in the Pro/ENGINEER family of modules. Windows, Windows 95, Windows NT, and Notepad are registered trademarks of Microsoft Corporation.

ABOUT THE AUTHOR

David S. Kelley is Assistant Professor of Manufacturing Graphics in the Department of Computer Graphics Technology at Purdue University. Prior to joining Purdue's faculty, David served as an assistant professor in the Engineering Technology Department at Western Washington University. He has also taught engineering graphics at Itawamba Community College in Fulton, Mississippi; drafting and design technology at Northwest Mississippi Community College; engineering design at Northeastern State University in Tahlequah, Oklahoma; and engineering graphics technology at Oklahoma State University—Okmulgee. He earned his A.A. degree from Meridian Community College in Meridian, Mississippi, his B.S. and M.S. degrees from the University of Southern Mississippi, and his Ph.D. degree from Mississippi State University. He may be reached at dskelley@tech.purdue.edu.

CONTENTS

CHAPTER

1

PRO/ENGINEER'S USER INTERFACE

Introduction

Pro/ENGINEER has both a UNIX and Windows version (NT, 95, and 98). Starting with Release 20, Pro/ENGINEER introduced a new interface. This interface resembles, on its face, a typical Windows application. When manipulating Pro/ENGINEER, it is important to remember that it does not function like a true Windows application. This chapter will introduce the fundamentals of Pro/ENGINEER's interface. Upon finishing this chapter, you will be able to

- Describe the purpose behind each menu on Pro/ENGINEER's menu bar.
- Use Pro/ENGINEER's file management capabilities to save object files.
- Set up a Pro/ENGINEER object to include units, tolerances, and materials.
- Customize Pro/ENGINEER through the use of the configuration file.
- Customize Pro/ENGINEER commands using mapkeys.
- Organize items using the Layers option.

DEFINITIONS

Configuration file   A Pro/ENGINEER file used to customize environmental and global settings. Configuration options can be set through the Utilities >> Preferences option.

Mapkeys   Keyboard macros used to define frequently used command sequences.

Model   An object that represents the actual sculptured part, assembly, or work piece.

Nominal dimension   A dimension with no tolerance.

Object   A file representing an item, part, assembly, drawing, layout, or diagram created in Pro/ENGINEER.

Tolerance   The allowable amount that a feature's size or location may vary.

MENU BAR

The following describes many of the options available on Pro/ENGINEER's Part mode menu bar (see Figure 1–1). While this interface may appear to make Pro/ENGINEER a true Windows application, many typical Windows functions are not available (Copy, Paste, etc.).

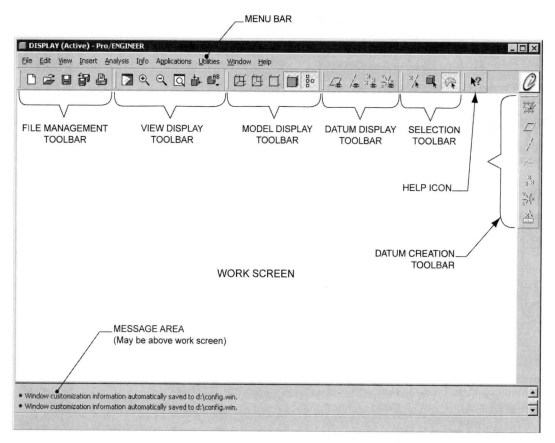

Figure 1-1 Pro/ENGINEER's work screen

FILE MENU

The File menu is Pro/ENGINEER's interface for the manipulation of files and objects. Found under the File menu are options for saving and opening objects. Also, options are available for printing and exporting objects.

EDIT MENU

The Edit menu provides options for the modification of geometric elements. Within part mode, options are available for performing feature manipulation and modification techniques such as Redefine, Reroute, Suppress, and Delete. Within sketch mode, options are available for moving, copying, and trimming sketched entities.

VIEW MENU

The View menu is used to change the appearance of models and Pro/ENGINEER's work screen. Many of the view options available exist as shortcut keys or can be found on the Toolbar. Commonly used view manipulation options under the View menu include zooming, repainting the view, and retrieving the default view. Options also are available for orienting a model, saving a view, modifying a model's color and appearance, and for changing the lighting of Pro/ENGINEER's work screen.

INSERT MENU

The Insert menu provides selections for the creation of traditional Pro/ENGINEER features (e.g., Protrusion, Hole, Datum Plane, Cosmetic Thread, etc.). Unlike previous versions of Pro/ENGINEER, datums can be created at anytime during object modeling, including while sketching.

ANALYSIS

Options for finding assembly and part properties can be found under the Analysis menu. As an example, the mass of a part can be obtained through the Model Analysis option.

INFO MENU

The Info menu provides information about Pro/ENGINEER objects. Information can be found on Parent-Child relationships, features, references, and geometry. Messages, such as error messages created during regeneration failures, can be displayed using the Message Log option. Additional information about failed regenerations can be found using the Geometry Check option. A commonly used option under the Info menu is Switch Dims, in which dimensions can be displayed with numeric values or with dimension symbols. This option toggles between the two dimension display modes.

APPLICATIONS MENU

The Applications menu will allow a user to switch between Pro/ENGINEER modes and applications. As an example, a user may switch between Part mode and Manufacturing mode.

UTILITIES MENU

The Utilities menu allows for the customization of Pro/ENGINEER's interface. Before Release 20, the Environment option was a common tool for changing the work screen appearance. Many of the features found under the environment menu, such as datum and model display, can now be found on the Toolbar. The Utilities menu provides access to Pro/ENGINEER's configuration file. Additional options are available for customizing colors and for creating **mapkeys.**

WINDOW MENU

The Window menu is used to manipulate Pro/ENGINEER windows. Windows can be activated, opened, or closed. Within Pro/ENGINEER, multiple windows of multiple parts can be open at once. To work in one menu, a user has to first activate it. Opened windows are displayed under the Window menu, thus allowing a user to easily switch from one object to another.

HELP MENU

Pro/ENGINEER utilizes a web browser to access help information. The Pro/Help CD must be loaded before the full help option can be utilized. While Pro/Help provides search capabilities for Pro/ENGINEER options, a context-sensitive help option is also available. Use context-sensitive help to find information on individual Pro/ENGINEER menus and options.

TOOLBAR

As shown previously in Figure 1–1, Pro/ENGINEER provides a toolbar for easy access to frequently used options through the use of icons. By default, Pro/ENGINEER's initial toolbar is divided into five groups. Additional options can be added to the toolbar under the Utilities menu.

FILE MANAGEMENT

The file management group of icons is available for manipulating files. These icons (see Figure 1–2) are

Figure 1-2   File management icons

- **New** The New icon is used to start a new Pro/ENGINEER file.
- **Open** The Open icon is used to open a Pro/ENGINEER file.
- **Save** The Save icon is used to save a Pro/ENGINEER file.
- **Save A Copy** The Save A Copy icon is used to save a Pro/ENGINEER file as a different name and to a different location.
- **Print** The Print icon is used to print or plot a Pro/ENGINEER object.

VIEW DISPLAY

The view display group of icons is available for modifying the display of Pro/ENGINEER objects on the work screen. These icons (see Figure 1–3) are:

Figure 1-3   View display icons

- **Repaint** The Repaint icon is used to redraw the work screen.
- **Zoom In** The Zoom In icon is used to zoom in to a user-defined window.
- **Zoom Out** The Zoom Out icon is used to zoom out from the work screen.
- **Refit** The Refit icon is used to fit the extent of a Pro/ENGINEER object into the work screen.
- **Orient** The Orient icon is used to orient a Pro/ENGINEER object on the work screen.
- **Saved Views** The Saved Views icon is used to access saved views.

MODEL DISPLAY

The model display group of icons is available for changing the display of Pro/ENGI-NEER objects. Only one of the four available icons under this group may be activated at a time. These icons (see Figure 1–4) comprise the following:

Figure 1-4   Model display icons

- **Wireframe** The Wireframe icon displays a Pro/ENGINEER object as a wireframe.
- **Hidden Line** The Hidden Line icon displays a Pro/ENGINEER object with hidden lines.

- **No Hidden** The No Hidden icon displays a Pro/ENGINEER object without hidden lines.
- **Shade** The Shade icon shades a Pro/ENGINEER object.
- **Model Tree** The model tree icon turns the display of the model tree on or off.

DATUM DISPLAY

The datum display group of icons is used to control the display of datums. This group (see Figure 1–5) contains the following:

Figure 1-5 Datum display icons

- **Datum Planes** The Datum Plane icon is used to turn on or off the display of Pro/ENGINEER datum planes.
- **Datum Axes** The Datum Axes icon is used to turn on or off the display of Pro/ENGINEER datum axes.
- **Point Symbols** The Point Symbols icon is used to turn on or off the display of Pro/ENGINEER datum points.
- **Coordinate Systems** The Coordinate Systems icon is used to turn on or off the display of Pro/ENGINEER coordinate systems.

CONTEXT-SENSITIVE HELP

The Context-Sensitive Help icon is used to display help information on individual menu or dialog box options. To use this help function, select the context-sensitive help icon (see Figure 1–6); then select the menu item. Pro/Help will launch a web browser and display information about the selected item.

Figure 1-6 Context-sensitive help

FILE MANAGEMENT

Various options are available for manipulating Pro/ENGINEER files. Pro/ENGINEER's file management capabilities provide a wide range of functions for managing projects and models. On first appearance, Pro/ENGINEER's file open and save commands resemble a Windows application. However, there are some significant differences between Pro/ENGINEER's file management and a Windows application.

- Pro/ENGINEER filename requirements are more restricted than Windows application filenames.
- Saving a Pro/ENGINEER object creates a new version of the object each time the object is saved. It does not override older versions.
- Pro/ENGINEER will not allow an object to be saved to a specific filename if that filename already exists. Pro/ENGINEER will not save on top of an existing file.

FILENAMES

Pro/ENGINEER has different file extensions according to the mode being utilized. Table 1–1 shows file extensions based on five common Pro/ENGINEER modes:

Table 1-1   File extensions for Pro/ENGINEER modes

Mode	Extension
Sketch	*.sec.*
Part	*.prt.*
Assembly	*.asm.*
Manufacturing	*.mfg.*
Drawing	*.drw.*
Format	*.frm.*

Notice the extra asterisk at the end of each file extension. This asterisk represents the version of the file. The first time Pro/ENGINEER saves a file, this extra extension has a value of 1. The second time a file is saved, a new file is created with a 2 as this value. The third time a file is saved, a new file is created with a 3 as this value. Pro/ENGINEER creates a new object file each time a file is saved. If an object file is saved 10 times, 10 Pro/ENGINEER files will be created. To delete the previous Pro/ENGINEER files, select File >> Delete >> Old Versions.

Pro/ENGINEER file and directory names cannot be longer than 31 characters. Brackets, parentheses, periods, nonalphanumeric characters, and spaces cannot be used in a filename. An underscore (_) may be used in a file name, though. Table 1–2 shows examples of invalid and valid filenames.

Table 1-2   Invalid and valid filenames

Invalid Filename	Problem	Valid Filename
part one	Space in filename	part_one
part@11	Nonalphanumeric character	part_11
Part[1_10]	Brackets used in filename	Part_1_10

MEMORY

When an object is opened, referenced, or created in Pro/ENGINEER, it is placed in memory. It remains there until it is erased, or until Pro/ENGINEER is exited. Also, when opening an assembly, every part referenced by the assembly is placed in memory. Parts in active memory are displayed in a window. Multiple parts, assemblies, and drawings can be in active memory at once. This allows for ease of access between objects. Objects may also be in session memory. Session memory is the condition where the object is in memory, but not displayed in a graphics window.

WORKING DIRECTORY

Pro/ENGINEER utilizes a Working Directory to help manage files. The working directory is usually the modeling point for all Pro/ENGINEER objects. When a new file is saved, it is saved in the current working directory, unless a new directory is specified. To change the current working directory, from the File menu select Working Directory; then select the desired directory as the working directory.

SAVING AN OBJECT

Various options are available for saving objects. New objects are saved by default in the current working directory. If an object is retrieved from a directory other than the working directory, the object is saved in its original directory. Additionally, selecting Save while in a sketcher environment will save the section (*.*sec.**) and not the object being modeled. The following options are available for saving objects.

SAVE

This option saves an object to disk. When saving an assembly, all individual parts that comprise the assembly are saved. When saving a drawing, the model used to create the drawing is saved only when changes have been made to the object. While sketching in a sketcher environment, the section under modification or creation is saved, but not the object file. Pro/ENGINEER objects can be saved to a computer hard drive, floppy disk, or zip disk.

SAVE A COPY

The Save A Copy option is used to either save an object as a new filename or to save an object to a new directory. When an object is saved using this option, the original filename is not deleted and is still the active model. Save A Copy practically creates a copy of the object being modeled. Any changes made to the original object are not reflected in the copied object.

BACKUP

The Backup option creates a copy of the object being modeled. The name of the object cannot be changed with this option. Any saves of an object conducted after a backup will be to the directory of the backup.

RENAME

The Rename option changes the name of a Pro/ENGINEER object. A suboption is available for renaming the object on disk and in memory, or just in memory. When renaming an object that already exists, all previous versions of the object are saved.

DELETE

Saving an object multiple times can create many versions of the object on disk. The Delete option is available to purge old versions. Options are available for deleting old versions or all versions of an existing object.

ERASE

Closing a window that contains a Pro/ENGINEER object does not remove it from memory. The Erase command must be used to remove an object from memory. An object that is referenced by another opened object cannot be erased. The Erase dialog box shows all objects referenced by a selected object. Options are available for erasing referenced objects from memory or for keeping them in memory.

ACTIVATING AN OBJECT

Multiple objects can be open at once within Pro/ENGINEER. Additionally, multiple windows can be opened. To modify an object, its associated window must be activated. To make a window active, use the Activate option found under the Window menu.

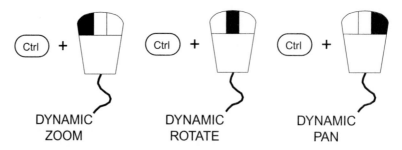

Figure 1-7   Dynamic viewing options

VIEWING MODELS

There are many different ways to view a Pro/ENGINEER object and to view Pro/ENGI-NEER's work screen. Options are available for panning, rotating, and zooming an object dynamically. Other options are available for changing the display of a model.

DYNAMIC VIEWING

A useful feature of Pro/ENGINEER that enhances its model building capabilities is its dynamic viewing functions. A model can be dynamically zoomed, rotated, and panned using a combination of the Control key and a mouse button (see Figure 1–7).

DYNAMIC ZOOM

A user can dynamically Zoom in or out on a model by using Pro/ENGINEER's dynamic zoom option. Dynamic zoom is activated when the keyboard's Control key is selected at the same time as the left mouse button. While simultaneously holding down the Control key and the left mouse button, moving the cursor from the top of the work screen to the bottom will zoom in on a model. Correspondingly, moving the cursor from the bottom of the work screen to the top will zoom out on a model.

DYNAMIC ROTATE

A user can dynamically Rotate a model by using Pro/ENGINEER's dynamic rotate option. Dynamic rotate is activated when the keyboard's Control key is selected at the same time as the middle mouse button. While simultaneously holding down the Control key and the middle mouse button, moving the cursor around the work screen will rotate the model around a specified spin center.

DYNAMIC PAN

A user can dynamically Pan a model by using Pro/ENGINEER's dynamic pan option. Dynamic pan is activated when the keyboard's Control key is selected at the same time as the right mouse button. While simultaneously holding down the Control key and the right mouse button, moving the cursor around the work screen will pan the model.

MODEL DISPLAY

As shown in Figure 1–8, four styles are used to display a model in part, assembly, and manufacturing modes. Similarly, within drawing mode, three styles are used. There are situations when each display style is the most practical. The Model Display dialog box is located under the View menu and contains other display options. Each display style can be selected dynamically from the Toolbar menu.

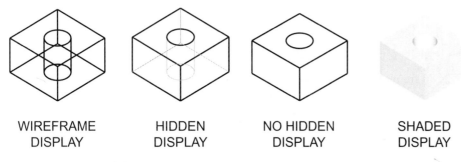

| WIREFRAME | HIDDEN | NO HIDDEN | SHADED |
| DISPLAY | DISPLAY | DISPLAY | DISPLAY |

Figure 1-8   Model display options

WIREFRAME DISPLAY

Within all relevant modes of Pro/ENGINEER, the Wireframe style displays all edges of a model as a wireframe. Edges that would be hidden from view during a true representation of the model are displayed, as are edges that would not be hidden.

HIDDEN DISPLAY

With the Hidden display style, lines that would be hidden from view during a true representation of a model are shown in gray. Within Drawing mode, these gray lines represent hidden lines and will be printed as hidden lines.

NO HIDDEN DISPLAY

With the No Hidden display style, lines that would be hidden from view during a true representation of a model are not shown.

SHADED DISPLAY

With the Shaded display, all solids and surfaces are displayed shaded. Hidden lines are not shown. This option is not available in drawing mode.

Three possible default views are available within Pro/ENGINEER: Trimetric, Isometric, and User-Defined. When selecting Default View from the View menu, the model returns to this viewpoint. The initial setting within Pro/ENGINEER is Trimetric. This setting can be permanently changed by the configuration file option *orientation,* or temporarily changed in the Environment menu or in the Orientation dialog box.

UNITS

Within Pro/ENGINEER, four Principle Unit categories exist: length, mass or force, time, and temperature. Each category has a full range of possible units. As an example, available within the length category are inches, feet, millimeters, centimeters, and meters.

SETTING A SYSTEM OF UNITS

Pro/ENGINEER utilizes a system of units to group the four principle categories. There are six predefined systems of units available within Pro/ENGINEER. These systems may be accessed through the System of Units tab found on the Units Manager dialog box.

- Meter kilogram second (MKS)
- Centimeter gram second (CGS)
- Millimeter newton second (mmNs)

- Foot pound second (FPS)
- Inch pound second (IPS)
- Inch lbm second (Pro/E Default)

As shown from the preceding list of predefined systems of units, *inch-lbm-second* is the default system. To set a different system of units, select the specific system on the System of Units tab, then select the Set button.

CREATING A SYSTEM OF UNITS

A user-defined system of units can be created to allow a user to establish units that meet design intent for a given product. Perform the following steps to create a new system of units.

STEP 1: **Select SET UP >> UNITS on Pro/ENGINEER's Menu Manager.**

When selecting the Units option, the Units Manager dialog box will appear (Fig. 1–9). Options available under this dialog box include: (*a*) creating a new system, (*b*) copying an existing system to create a new system, (*c*) reviewing a system's units, (*d*) deleting a user-defined system, and (*e*) setting a system of units.

STEP 2: **Select NEW from the Units Manager dialog box.**

STEP 3: **In the System of Units Definitions dialog box, enter a NAME for the user-defined system (Figure 1–9).**

STEP 4: **Select UNITS from each principle category.**

Notice under the dialog box that an option exists for choosing between Mass and Force. Each subcategory has its own available units.

STEP 5: **Select OKAY, then close the Units Manager dialog box.**

MODELING POINT   It is good modeling practice to set your units before creating a feature. If you forget to set them before creating the first feature, units can be converted from one system to another. Pro/ENGINEER will interpret your existing dimensions as another defined unit (e.g., 1" will be interpreted as 1mm). To allow for this interpretation, use the Interpret Existing Numbers option after setting a new unit.

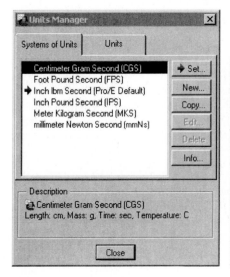

Figure 1-9   Units menus

PRINTING IN PRO/ENGINEER

Pro/ENGINEER has the capability to print to a variety of printers and plotters. Additionally, objects may be printed to a file. The Print dialog box, as shown in Figure 1–10, can be opened by selecting Print from the File menu or from the Toolbar.

The Print Dialog box Destination is used to select a specific printer. To change printers, select the Add Printer icon.

> **MODELING POINT** The Windows Printer Manager can be used to print objects. To use the default Windows printer, in the configuration file, enter *Windows Printer Manager* as the value for the Plotter option.

CONFIGURING THE PRINTER

To configure a printer, select the Configure button located on the Printer dialog box. Multiple print configurations can be set. To save a configuration, select the Save button. The following tab options are available under Configure.

PAGE TAB

As shown in Figure 1–11, the Page tab is used to configure the sheet size. Standard sheet sizes are available (A, B, C, D, E, F, A0, A1, A2, A3, and A4), or a user may specify a variable sheet size.

PRINTER TAB

As shown in Figure 1–12, the Printer tab is used to specify printer options that might be available. Not all options are available with every printer.

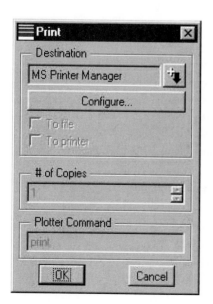

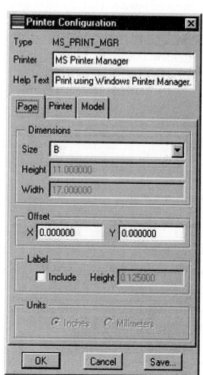

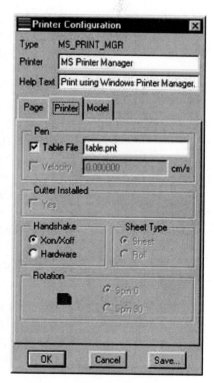

Figure 1–10 Print dialog box   **Figure 1–11** Page tab   **Figure 1–12** Printer tab

MODEL TAB

The Model tab is used to adjust the way an object is printed on a sheet. The Plot field is used to adjust the area to be plotted. Options are available for creating a Full Plot, a Clipped plot, a plot based on Zoom, a plot of an Area, and a plot based on the Model Size. The Base-on-Zoom option is the default value. For full sized plots, this should be changed to Full Plot.

PRO/ENGINEER'S ENVIRONMENT

Various selections are available for controlling Pro/ENGINEER's working environment. Figure 1–13 shows Pro/ENGINEER's Environment dialog box. This dialog box is available under the Utilities menu. Many of the Environment selection options are also available through other avenues, such as the Toolbar menu. The following is a description of each available selection.

- **Dimension Tolerances** This selection specifies whether an object's dimensions are displayed as a tolerance or as a nominal value. This option is also controllable with the configuration file option *Tolerance Display*. A dimension's tolerance mode can be set with the Modify >> Dimension option, or by default with the configuration file option *Tolerance Mode*.
- **Datum Planes** This selection option controls the display of datum planes. This option is readily accessible through Pro/ENGINEER's Toolbar menu.
- **Datum Axes** This selection option controls the display of datum axes. This option is readily accessible through Pro/ENGINEER's Toolbar menu.
- **Point Symbols** This selection option controls the display of datum points. This option is readily accessible through Pro/ENGINEER's Toolbar menu.
- **Coordinate Systems** This selection option controls the display of coordinate systems. This option is readily accessible through Pro/ENGINEER's Toolbar menu.
- **Spin Center** The Spin Center is used to show the center of a spin when an object is rotated. This option is activated by default. This selection can be changed permanently using the configuration file option *spin_center_display*.
- **3D Notes** This option controls the display of 3D notes while in part or assembly mode.
- **Thick Cables** This selection controls the three-dimensional display of cables. The Centerline Cables option and this option cannot be selected at the same time.
- **Centerline Cables** This selection displays the centerlines of cables. The Thick Cables option and this option cannot be selected at the same time.
- **Internal Cable Portions** This selection option allows for the display of cables that are hidden by other geometry.
- **Model Tree** This option controls the display of Pro/ENGINEER's Model Tree.
- **Colors** This option will display a model in color.
- **Levels of Detail** This selection will allow for the use of Levels-of-Detail while dynamically viewing a shaded model.
- **Ring Message Bell** When selected, a bell will sound when Pro/ENGINEER provides a message. This selection can be changed permanently with the configuration file *Bell* option.
- **Save Display** This option will save the display of an object, reducing the recalculations needed when reopening the object at a future time.

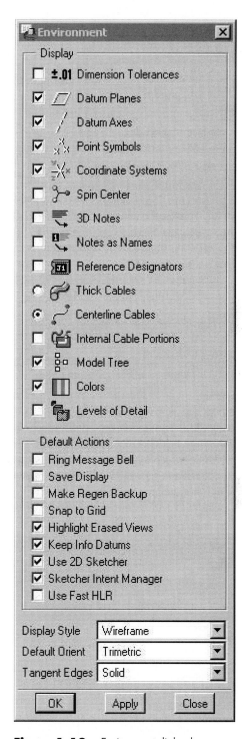

Figure 1-13   Environment dialog box

- **Make Regen Backup** This selection creates a backup copy of the object before regenerations. This allows for the retrieval of a previously valid model. Backups are deleted when ending the object's session.

- **Snap to Grid** When Grid is activated, elements will snap to them, particularly within the sketcher environment.

- **Keep Info Datums** This selection option controls the display of datums created on-the-fly with the Info functionality. When this option is selected, these datums will be considered as features within the model.

- **Use 2D Sketcher** By default, sketching is set up within a two-dimensional environment. Using this setting, the sketching plane is oriented parallel to the screen. When not selected, the sketcher environment remains in a three-dimensional orientation.

- **Use Fast HLR** This option allows for the faster acceleration of hardware while dynamically viewing a model.

- **Display Style** There are four display styles available for viewing a model: Wireframe, Hidden Line, No Hidden, and Shading. These options are also available on Pro/ENGINEER's Toolbar menu.

- **Default Orient** There are three settings available for Pro/ENGINEER's default orientation: Isomeric, Trimetric, and User Defined. Trimetric is the initial setting. This can be changed with the configuration file option *Orientation*.

- **Tangent Edges** This selection option controls the display of tangent edges. There are five options available: Solid, No Display, Phantom, Centerline, and Dimmed (Dimmed Menu Color).

CONFIGURATION FILE

Pro/ENGINEER's Options dialog box (Figure 1–14) is used to customize a variety of environmental and global settings. It is accessed through the Utilities >> Options command. Options such as model orientation, system geometry color, background color, tolerance mode, and sketcher grid display can be set by default. As shown in the figure, the left-most column of the table is used to establish a specific option, and the next column is used to define the value for the option. Available options can be found through Pro/ENGINEER's online help. A user can set an option by entering the option's name and value (e.g., *orientation* and *isometric* as shown in the illustration), followed by selecting the Add/Change icon.

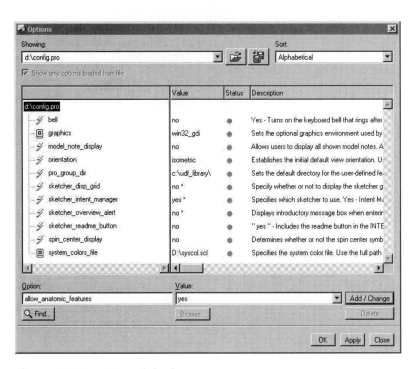

Figure 1-14 Options dialog box

MODELING POINT The configuration file is used to permanently set environmental and global settings. Most settings can be changed temporarily using other options, such as under the Environment dialog box.

Configuration files may be defined in a variety of directories. When Pro/ENGINEER is first launched, it reads configuration files in order from the locations listed below. The last settings read from a configuration file are the ones that Pro/ENGINEER utilizes.

1. **Pro/ENGINEER LOADPOINT** The loadpoint directory is the location where Pro/ENGINEER is installed. A configuration file saved here is loaded first.
2. **LOGIN DIRECTORY** The login directory is the home directory for a login ID. This configuration file is read after the loadpoint directory and is used to save individualized configuration options.
3. **STARTUP DIRECTORY** The startup directory is the working directory for the current object. This is the last configuration file read by Pro/ENGINEER and any settings that are read from this file will override settings from previous configuration files.

LAYERS

Layers are used to organize features and parts together to allow for the collective manipulation of all included items. Features and parts can be included on more than one layer. The Layers dialog box is used to create and manipulate layers (Figure 1–15). It is accessed through the Layers option found under the View menu. The New Layer icon, found on the Layers dialog box, is used to create new layers. The dialog box has options for setting features and parts to selected layers or to remove items from a layer. It is used to control how layers are displayed in a model. The following options are available.

- **Show** Used to display a layer.
- **Blank** Used to blank the display of a layer. Items will not be displayed if their associated layers are blanked.

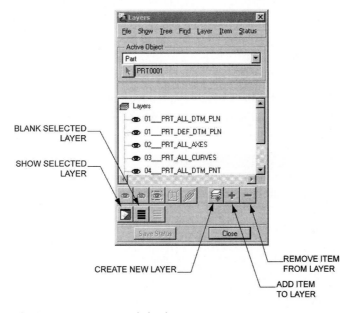

Figure 1-15 Layers dialog box

- **Isolate** Layers that are selected for isolation will be shown while unselected layers and items not on a layer will be blanked.
- **Hidden** The Hidden option is available in Assembly mode only. Parts in hidden layers will not be displayed according to environmental settings for hidden-line display.
- **Add Items** The Add Items option is used to add selected items to a layer.
- **Remove Items** The Remove Items option is used to remove items from layers.

CREATING A LAYER

Perform the following steps to create a new layer:

STEP 1: Select VIEW >> LAYERS on Pro/ENGINEER's Menu Bar.

The Layers dialog box will appear after selecting the Layers option (Figure 1–15).

STEP 2: On the Layers dialog box, select the CREATE NEW LAYER icon (Fig. 1–15).

STEP 3: On the New Layer dialog box, enter a Name for the new layer (Fig. 1–16).

STEP 4: Select OKAY on the New Layer dialog box to create the new layer.

Selecting OKAY will create the new layer and return you to the Layers dialog box. Selecting the Return key on the keyboard will create the layer and allow you to remain in the New Layer dialog box. This allows for the creation of multiple layers.

SETTING ITEMS TO A LAYER

Perform the following steps to set an item or feature to a layer.

STEP 1: Select VIEW >> LAYERS on Pro/ENGINEER's menu bar.

STEP 2: On the Layers dialog box, pick an existing layer (Figure 1–15).

STEP 3: On the Layers dialog box, pick the ADD ITEM icon (Figure 1–15).

After selecting the Add Item icon, Pro/ENGINEER will launch the Layer Object menu. This menu is used to select an item type to add to the layer.

STEP 4: On the Layer Object menu, select an item type to add to the layer.

Examples of items that can be added to a layer include features, curves, and quilts.

STEP 5: On the work screen or on the model tree, select items to add to the layer.

STEP 6: On the Layer Feature menu, select DONE/RETURN.

STEP 7: On the Layer Object menu, select DONE/RETURN.

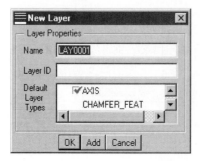

Figure 1-16   New Layer dialog box

Summary

Pro/ENGINEER provides a variety of tools for interfacing with the modeling environment. Most of these tools are readily available on the toolbar. Other options can be found under the menu bar.

Pro/ENGINEER's data-based management system manipulates files differently than standard Windows applications. Options are available for saving and backing up object files. A unique feature of Pro/ENGINEER is that when it saves, a new version of the object file is created.

A variety of view manipulation options is available within Pro/ENGINEER. Objects can be dynamically zoomed, rotated, and panned. Views can be oriented and saved for later use. Additionally, the display of an object can be represented in one of four possible ways: wireframe, hidden line, no hidden, and shaded. Also, features and entities may be placed on layers and hidden from display.

Pro/ENGINEER's interface provides customization tools. The configuration file is a powerful tool for personalizing the work environment. Mapkeys can be created that provide a shortcut to commonly used menu pick sequences.

Problems

1. Create a system of units with the following configuration:

    *   Length = cm
    *   Mass = kg
    *   Time = micro-sec
    *   Temperature = K

2. Create a configuration file with the following options and settings:

    *   *BELL* option set to NO.
    *   *TOL_DISPLAY* option set to YES.
    *   *TOL_MODE* option set to NOMINAL.
    *   *SKETCHER_INTENT_MANAGER* option set to YES.
    *   *ALLOW_ANATOMIC_FEATURES* option set to YES.

Questions and Discussion

1. Describe the difference between the Backup option and the Save A Copy option. Give some examples of when each option would be appropriate.

2. In the object filename *revolve.sec.2,* what does the number *2* represent? What does *sec* represent?

3. What is a Pro/ENGINEER working directory?

4. What file management option is used to close an object from Pro/ENGINEER's memory?

5. What is the purpose of Pro/ENGINEER's configuration file?

2

EXTRUDING, MODIFYING, AND REDEFINING FEATURES

Introduction

This chapter introduces extruded features, a concept associated with basic modeling fundamentals. Within Pro/ENGINEER, the Extrude option is common among the Protrusion and Cut commands. Additionally, this chapter will introduce the Redefine command, feature modification techniques, and datum construction. Upon finishing this chapter, you will be able to

- Model solid features as extruded protrusions.
- Remove material from features using extruded cuts.
- Modify feature dimension values using the Modify command.
- Modify feature definitions using the Redefine command.
- Create datum planes.

DEFINITIONS

Base feature   The first geometric feature created in a part. It is the parent feature for all other features.

Child feature   A feature whose definition is partially or completely referenced to other part features. A feature referenced by a child feature becomes a parent of this feature.

Cut   A negative space feature created from a sketched section.

Definition   A parameter of a part. An example of a definition of a hole feature would be the depth of the hole.

Protrusion   A positive space feature created from a sketched section.

Negative space feature   A feature created by removing material from a model. Examples of negative space features include holes, cuts, and slots.

Parent feature   A feature referenced by another feature.

Positive space feature   A feature created by adding material to a model. Examples of positive space features include protrusions and ribs.

FEATURE-BASED MODELING

Parametric design packages are often referred to as feature-based modelers. A *feature* is a subcomponent of a part that has its own parameters, references, and geometry. *Geometry*

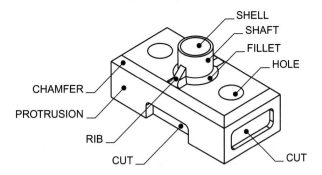

Figure 2-1   Features of a part

(see Figure 2–1) is the graphic description of a feature. Geometry can be sketch-defined or predefined. Sketch-defined features consist of sketched sections that are protruded or cut to form either positive or negative space. Predefined geometry has a common section such as a hole, round, or chamfer. Parameters are the dimensional values and definitions that define a feature. A hole may have a diameter of 1 inch and can be extruded completely through all existing features. The diameter is a parameter, as is the through-all definition. Parametric modeling packages allow users to modify parameters after the feature has been modeled. This is one of the unique properties that separates parametric modelers from Boolean-based modelers. *References* are ways that features are related to other features in a part or assembly. Examples of references include axes, sketch planes, placement planes, reference planes, and reference edges. The surface of one feature may serve as the sketch plane for a second feature. The edges of the first feature may also serve as reference lines for parameters defining the second feature. In both examples, the first feature is a parent of the second feature.

PARENT-CHILD RELATIONSHIPS

Parametric models are composed of features that have established relationships. Features build upon other features in a way that resembles a family tree, hence the phrase *parent-child relationship*. Actually, a history tree of the relationships between features in a Pro/ENGINEER model resembles a web. The first feature created in a part is the center of the web and is the **parent feature** for all features. **Child features** branch off from the base feature and themselves become parent features. Unlike a typical family tree, a child feature may have several parent features.

Parent-child relationships can be established between features implicitly or explicitly. Implicit relationships can be established through the adding of a numeric equation using the Relations option. An example of this would be making two dimensions of equal value. In this process, one dimension governs the value of another. The feature with the governing dimension is the parent feature of the feature with the governed dimension. Care should be taken when modifying a feature that has a dimension that governs another. If a parent feature is selected for deletion, Pro/ENGINEER will provide an error message requesting an action to be accomplished to satisfy the void relationship. The user has the option of deleting, modifying, redefining, or rerouting the relationship.

Explicit relationships are created when one feature is used to construct another. An example would be selecting a plane of one feature as the sketch plane for a second feature. The new feature will become a child of the feature being sketched upon. Another similar example of an explicit relationship would be using existing feature edges within the sketcher environment to create a new feature. By specifying references while sketching, these selected references will create a relationship between the feature being sketched and the existing feature being referenced. The new sketched feature becomes a child of any referenced feature.

PROTRUSIONS AND CUTS

The procedures for performing a Protrusion and Cut in Pro/ENGINEER are virtually identical. The primary difference between the Protrusion command and the Cut command is that a **protrusion** is a **positive space feature,** while a **cut** is a **negative space feature.** When you protrude a feature, you actually create a solid object. With the Cut command, an extruded feature removes material from existing features.

The menu structure for both commands is similar. For both, the following options exist:

EXTRUDE

The Extrude option sweeps a sketched section along a straight trajectory. The user draws the section in the sketcher environment and then provides an extrude depth. The section is protruded the depth entered by the user.

REVOLVE

The Revolve option sweeps a section around a centerline. The user sketches a profile of the revolved feature and a centerline to revolve about. The user then inputs the degrees of revolution.

SWEEP

The Sweep option protrudes a section along a user-sketched trajectory. The user sketches both the trajectory and the section.

BLEND

The Blend option joins two or more sketched sections. The trajectory may be straight or revolved.

USE QUILT

Quilts are patchworks of surfaces. The Use Quilt option turns a quilt into a solid feature.

ADVANCED

Common modeling options under the Advanced menu include Variable Section Sweep, Swept Blend, and Helical Sweep.

Shown in Figure 2–2 is an illustration of how one section can be used to create an Extrude, Revolve, Sweep, or Blend feature.

SOLID VERSUS THIN FEATURES

When creating a Protrusion or Cut, Pro/ENGINEER gives the option of choosing either a solid feature or a thin feature. Solid features are objects that are completely enclosed with material. Thin features are often confused with surfaces. In Pro/ENGINEER, surfaces are quilts with no defined thickness, whereas thin features are actually solids with a user-defined thickness. As shown in Figure 2–3, when the section is extruded as a solid, the section's feature is completely enclosed with material. When the section is extruded as a thin feature, the walls of the section are protruded with the provided wall thickness only.

Thin features can be used with all forms of the Extrude, Revolve, Sweep, and Blend options under the Protrusion and Cut commands. An example of an extruded thin cut is

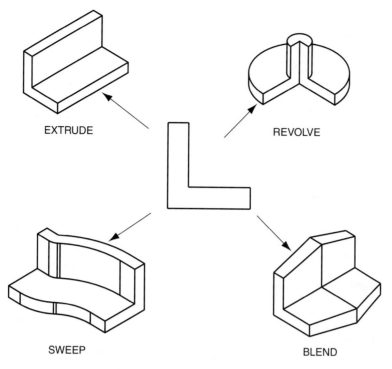

Figure 2-2   Variations in menu option features

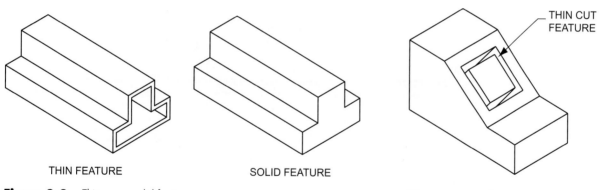

Figure 2-3   Thin versus solid feature   **Figure 2-4**   Thin cut feature

shown in Figure 2–4. The Thin option may be used with the Protrusion command for the base or secondary feature of a part or with the Cut command for secondary features.

EXTRUDED FEATURES

The following section will explore options available within extruded Protrusions and Cuts.

EXTRUDE DIRECTION

When sketching on a plane, Pro/ENGINEER, by default, specifies an extrude direction. When sketching on a datum plane, this direction is in the positive direction. When sketching on an existing feature, a Protrusion, as shown in Figure 2–5, will be extruded away

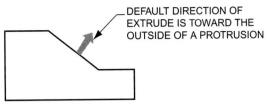

PROTRUSION DEFAULT EXTRUDE DIRECTION

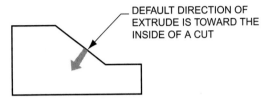

CUT DEFAULT EXTRUDE DIRECTION

Figure 2-5   Default extrude direction

from the feature. Since the objective of a Cut is to remove material, a Cut will be extruded toward the feature.

The Extrude option gives the user the option of flipping the direction of extrusion or specifying an extrusion in both directions. The Both Sides selection protrudes a section outward from the sketch plane in both directions. If the extrude depth is input to be 1.00 inch, the total extrusion will be 1 inch, not 1 inch in both directions. A typical Both Sides extrusion will divide the specified depth and extrude equally on both sides of the sketching plane. The 2 Side Blind depth option, though, allows the user to input unequal extrusion distances on both sides of the sketch plane.

> **MODELING POINT**   Features are composed of definitions established by the user. A definition is not permanently set. It can be changed before finishing a feature or with the Redefine command. If a definition such as the extrude direction, depth option, or material removal side is incorrectly set, do not cancel or delete the feature. Redefine it later.

DEPTH OPTIONS

For Extruded Protrusions and Cuts an important parameter is the distance of extrusion. Pro/ENGINEER provides eight basic ways to specify an extrusion's depth. Four common depth options are shown in Figure 2–6. The depth for an extrusion is entered for a feature after exiting the sketching environment.

BLIND

Blind is the simplest and most basic of the depth options. The Blind option allows a user to input an extrusion distance. It is the most common option for extruded base features.

2 SIDE BLIND

A 2 Side Blind is used with a Both Sides direction option only. This depth option will allow the user to enter separate extrude depths for both sides of the sketch plane.

THRU NEXT

The Thru Next option extrudes a feature to the next part surface. Part geometry must exist prior to using this option.

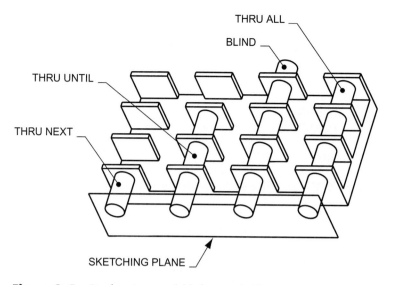

Figure 2-6 Depth options available for extruded features

THRU ALL

Thru All is one of the most common depth options for cut features. It extrudes a feature through the entirety of a part. The design intent of many material removal features (such as a Cut or Hole) is to cut completely through a part. Entering a blind depth that will extrude through the part may not be adequate if the part thickness changes. The Thru All option adjusts for changing part dimensions. This option is available for parts with existing features and is not available for surface features.

THRU UNTIL

The Thru Until option extrudes a feature until a user-selected surface. The surface can be any geometry, but cannot be a datum.

PNT/VTX

The Pnt/Vtx option extrudes a feature up to a selected datum point or vertex.

UPTO CURVE

The UpTo Curve option extrudes a feature up to a selected edge, axis, or datum curve.

UPTO SURFACE

The UpTo Surface option extrudes a feature up to a selected surface.

OPEN AND CLOSED SECTIONS IN EXTRUSIONS

Extruded sections may be sketched opened or closed. With the obvious exception of a base feature, many sections for an extruded Protrusion or Cut will suffice with an open section. The following are guidelines to follow when considering an open or closed section.

- Sections may not branch, and they can have only one loop. As shown in Figure 2–7, when sketching a section aligned with the edges of an existing feature, it often is not necessary to sketch over the existing geometry. Aligning the required sketch with the existing geometry will usually create a successful section. If Pro/ENGINEER is not sure which side of the section to protrude or cut, it will require the user to select a side (see Figure 2–8).

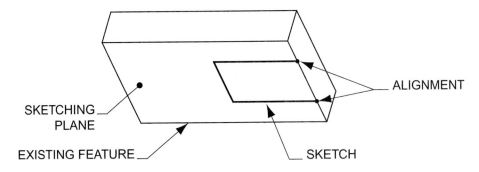

Figure 2-7   Aligning a sketch with existing part geometry

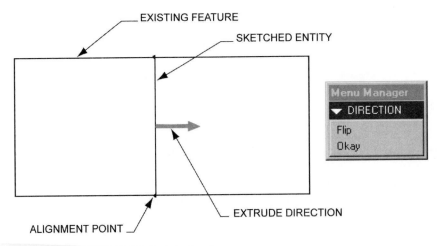

Figure 2-8   Selecting an extrude direction

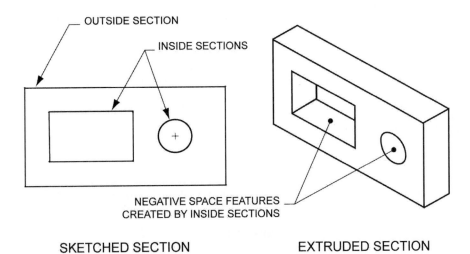

Figure 2-9   Creating negative space with included sections

- Thin feature sections may be open or closed.
- For thin features, sections can be open when not aligned with existing geometry.
- Multiple closed sections can be included in a sketch. As shown in Figure 2–9, when a section is included within another, the inside section creates negative space.

MATERIAL SIDE

Two definitions are associated with Material Side. The most common is the Material Removal Side. Within the Extrude option, the Material Removal Side definition is relevant only to the Cut command. The Material Removal Side definition is used to specify what side of a section that material will be removed from. By default, the removal side is toward the inside of a section. The user has the option of flipping the direction. Material Side definitions can be used with Protrusions when a sketch is an open section. Often, Pro/ENGINEER cannot determine which side of the sketch should be extruded. When this situation occurs, the user has to input the material side.

DATUM PLANES

Datum planes are used as references to construct features. Datum planes are considered features, but they are not considered model geometry. When a datum plane is created through the use of the Datum Toolbar, the datum plane will show as a feature on the model tree. A datum plane can be created and used as a sketch plane where no suitable one currently exists. As an example, a datum plane can be constructed tangent to a cylinder. This will provide a sketching environment that can be used to construct an extruded feature through the cylinder (Figure 2–10). When a feature is created on a datum plane, the datum plane is considered a parent feature.

A datum plane continues to infinity in all directions, with one side yellow and the opposite side red. Protrusions and orientations occur initially toward the yellow side of the datum plane. By default, datum planes are named in sequential order starting with DTM1. As a note, template files may have datum planes that have been renamed (e.g., Front, Top, and Right) with the Set up >> Name option.

Many times, a datum plane would be useful, but cluttering the model with additional features could be detrimental or confusing. One solution to this problem is to make a datum plane that is used for single feature creation only. To do this, datum planes can be created on-the-fly using the Make Datum option. Datum planes created on-the-fly belong to the feature undergoing creation. These datum planes do not show on the model tree and become invisible after the feature is created.

New to Release 2000i^2 of Pro/ENGINEER, a datum plane can be created at almost any point in the modeling process, including during the middle of a sketching environment. Traditionally, datum planes are created exclusively from other features. Starting with

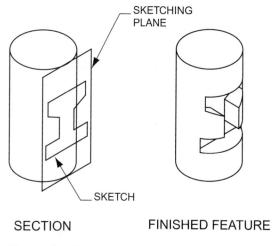

SECTION   FINISHED FEATURE

Figure 2-10   *Datum plane tangent to cylinder*

$2000i^2$, this new form of an on-the-fly datum still creates datum planes as features, but allows for flexibility during the creation process.

CREATING DATUM PLANES

A datum plane can be created at any point in the modeling process. One of the primary uses of a datum plane is as a sketching surface. Datum planes can be used as a mirror plane within the Copy command, or they can be used as references when sketching a feature. Datum planes are powerful features for aligning and mating parts within Assembly mode. Additionally, datum planes can be combined with other features using the Group option and then patterned.

Creating datum planes is a vital skill needed by all Pro/ENGINEER users. Several constraint options exist for datum plane definition. Some constraint options are stand-alone, while some are not. Paired constraint options are not stand-alone. They require two or more constraint definitions during the datum plane construction process. As an example, the Angle option requires the selection of an existing plane from which to reference the angle, then a Through constraint option to pass the datum plane through an axis or edge. Stand-alone constraint options require one option only.

STAND-ALONE CONSTRAINT OPTIONS

The following options only require one constraint to define a datum plane:

THROUGH >> PLANE

The Through >> Plane constraint option creates a datum plane that passes through an existing part plane.

OFFSET PLANE

The Offset Plane constraint option creates a datum plane that is offset from an existing plane. The user selects the plane from which to offset. Two offset options are available:

- **Point** Select a point to pass the datum plane through. The plane will be created through the point and parallel to the existing part plane.

- **Enter value** Enter an offset distance value. When selecting this option, the user is prompted to enter the offset value. An arrow in the graphics screen shows the default direction of the offset. To offset in the opposite direction, enter a negative value. The plane will be created offset from the reference plane at the value entered.

OFFSET/COORD SYS

The Offset Coordinate System constraint option creates a datum plane offset from the coordinate origin and normal to a selected coordinate axis. A coordinate system has to exist prior to the use of this option.

BLEND SECTION

The Blend Section constraint option creates a datum plane through a section used to create a feature.

PAIRED CONSTRAINT OPTIONS

The following options require two or more constraints to define a datum plane.

THROUGH>> AXISEDGECURV

This option is similar to the Through >> Plane option, except this option places a plane through an axis, edge, or curve. An axis, edge, or curve selected with the

Through option will not fully constrain a datum plane; hence, a second constraint option is required.

THROUGH >> POINT/VERTEX

This option places a datum plane through a point or vertex. Similar to the Through >> AxisEdgeCurv option, this option will not fully define a datum plane and needs an additional constraint option.

NORMAL >> AXISEDGECURV

This option places a datum plane perpendicular to an axis, edge, or curve. As with the Through >> AxisEdgeCurv option, an additional constraint is needed.

TANGENT >> CYLINDER

This constraint option places a datum plane tangent to a hole or cylindrical surface. This is an extremely useful option since it allows a feature to be constructed on the surface of a cylinder. This constraint option is often paired with the Normal >> Plane option or the Angle >> Plane option.

ANGLE >> PLANE

This option places a datum at an angle to an existing plane. It is often paired with the Through option, or the Tangent >> Cylinder option.

MODIFYING FEATURES

What separates parametric modeling packages, such as Pro/ENGINEER, from boolean-based modeling packages are their feature modification capabilities. Features created within Pro/ENGINEER are composed of parameters. Examples of parameters include parametric dimensions, extrude depth, and material side. Parameters such as these are established during feature construction. These feature parameters, and other feature definitions such as a section's sketch and sketch plane, can be modified later in the part modeling process.

As shown in Figure 2–11, a variety of feature modification options can be found under the Modify menu.

DIMENSION MODIFICATION

Parametric dimensions are used to define a feature. They can be modified at any time. Modifying a dimension value is the most common dimension modification function, but other modification tools exist. The number of decimal places in a dimension can be modified along with the tolerance format. The following are dimension modification techniques that are available within Pro/ENGINEER.

Figure 2-11   Modify options

MODIFYING A DIMENSION VALUE

The value of a parametric dimension value can be modified. To modify a dimension, select the Value option from the Modify menu, then select the feature associated with the dimension to be modified. Select the dimension to modify and enter a new value. Modifying a dimension value requires the regeneration of the part. Select Regeneration from the Part menu.

MODELING POINT   To display tolerances for all objects created within Pro/ENGINEER, the configuration file option *tolerance_mode* must be set to a value (such as Limit) and the option *tolerance_display* must be set to Yes. To display tolerances for individual objects, select Tolerance Display on the Environment dialog box.

TOLERANCE MODE
MODIFICATION

DECIMAL PLACE
MODIFICATION

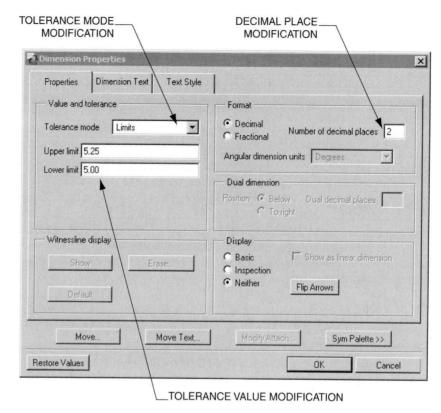

TOLERANCE VALUE MODIFICATION

Figure 2-12   Dimension Properties dialog box

MODIFYING A TOLERANCE MODE

Dimension tolerances can be displayed in a variety of modes. To modify the tolerance mode for individual dimensions, access the Dimension Properties dialog box by selecting Dimension under the Modify menu. As shown in Figure 2–12, an option exists on the Dimension Modification dialog box that allows for the changing of a tolerance mode.

MODIFYING TOLERANCE VALUES

The Value option under the Modify menu allows for the modification of tolerance values. As an example, if a dimension is set to Limits as the tolerance mode, either the upper or the lower dimension value can be changed. A problem with this approach is that the nominal value of the limit dimension cannot be modified. Another approach is to modify the nominal value and/or the tolerance values with the Dimension Properties dialog box (Figure 2–12). To access this dialog box, select Dimension from the Modify menu.

DIMENSION DECIMAL PLACES

Initially, dimension decimal places are set to two. This value can be changed permanently with the configuration file option *default_dec_places*. To change decimal places for individual dimensions, access the Dimension Modification dialog box by selecting Dimension from the Modify menu. As shown in Figure 2–12, an option exists for changing the decimal places of a dimension.

COSMETIC DIMENSION MODIFICATION

As shown in Figure 2–11, the DimCosmetics option under the Modify menu provides tools for modifying dimensions. A dimension's cosmetics are the way that the dimension appears on the object. Modifying a dimension's cosmetics does not modify the value of the

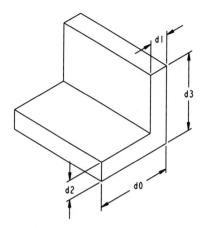

Figure 2-13   Dimension symbols

dimension. The following are techniques that the DimCosmetics option provides for modifying dimensions.

TOLERANCE FORMAT

A tolerance's format mode can be changed using the Format option found under the DimCosmetics menu. Select the dimensions for format change, then select the desired format to use.

NUMBER OF DIGITS OF A DIMENSION

The number of decimal places displayed by a dimension can be changed using the Num Digits option found under the DimCosmetics menu. To change the number of decimal places for dimensions, select Num Digits, then enter the number of significant digits. Follow this by selecting the dimensions to change.

ADDING TEXT AROUND A DIMENSION

Text can be added around a dimension value. A dimension's value is shown in a dimension note with the symbol @D. This symbol must remain in the dimension note. To add text around a dimension value, select Text from the DimCosmetrics option then select the dimension to modify. Follow this by entering lines of text.

CHANGING DIMENSION SYMBOLS

Dimensions within Pro/ENGINEER can be displayed in two ways. The first way is by showing the actual dimension or tolerance value. The second way is by showing the dimension symbol. As shown in Figure 2–13, when a dimension is created, it is provided with a dimension symbol. For the first dimension created on a part, the default symbol is d0. This number increases in sequential order for every new dimension. To change a dimension's symbol to make it more descriptive, utilize the Symbol option from the DimCosmetrics menu.

REDEFINING FEATURES

Features are composed of parameters. Parameters can be modified and changed through the Redefine command or through the Modify menu. Many different varieties of parameters exist. The following is a partial list.

- A feature's section.
- The sketch plane for a feature.
- The depth option for a feature.

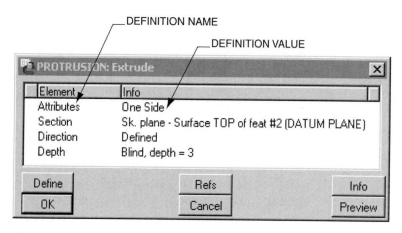

Figure 2-14   Feature Definition dialog box

- The material removal side.
- The extrude direction of a feature.
- The trajectory of a swept feature.
- The value of a dimension.

Figure 2–14 shows an example of a Feature Definition dialog box for an extruded Protrusion. This dialog box is accessible at the end of the construction of a part feature or it can be accessed through the Feature menu's Redefine command. Perform the following steps to redefine a feature parameter.

STEP 1:   **Select FEATURE >> REDEFINE.**

STEP 2:   **On the work screen or on the Model Tree, select the feature to be redefined.**

After selecting the feature to redefine, the Feature Definition dialog box associated with the feature will appear (Figure 2–14). This dialog box is composed of two columns. The first column displays the definition's name, while the second column displays the definition's current value.

STEP 3:   **On the Feature Definition dialog box, select the element to redefine.**

Select the definition's name displayed in the first column of the dialog box.

STEP 4:   **Select DEFINE on the dialog box.**

After selecting Define, Pro/ENGINEER will step you through the remodeling of the parameter.

STEP 5:   **Redefine the parameter according to Pro/ENGINEER modeling procedures.**

STEP 6:   **View the feature's new parameters by selecting PREVIEW on the dialog box.**

STEP 7:   **Select OKAY on the dialog box.**

SUMMARY

Pro/ENGINEER is often found difficult to use. Part of the reason is the menu intensity required to create a feature. It is true that several menu selections are required to create a feature, but many of these selections are common throughout different types of features. As an example, the Protrusion and Cut commands have basically the same menu structure and suboptions. One of the keys to learning Pro/ENGINEER is to become familiar with these options and to adapt them to new modeling situations.

EXTRUDE TUTORIAL

This tutorial exercise provides step-by-step instruction on how to model the part shown in Figure 2–15.

This tutorial will cover:

- Starting a new model.
- Setting default datum planes.
- Creating an extruded protrusion.
- Creating an extruded cut.
- Dimension modification.
- Redefining a feature.
- Saving a part.

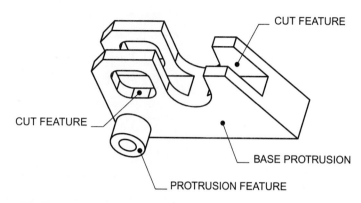

Figure 2-15   Finished part

STARTING A NEW MODEL

This segment of the tutorial will explore the starting of a new part file.

STEP 1: **Start Pro/ENGINEER.**

STEP 2: **Use the FILE >> SET WORKING DIRECTION option to establish a working directory for your part file.**

Pro/ENGINEER utilizes a Working Directory to help manage files. When a new file is saved, it is saved in the current working directory, unless a new directory is specified. When the Open command is executed, the default directory is the current working directory.

STEP 3: **Use the FILE >> NEW option to create a new part file.**

From the File menu, select the New option. Part is one of the modes of Pro/ENGINEER, and is the default mode. Enter a name for the new part. Part names must be less than 31 characters and cannot include spaces.

Notice on your New dialog box how the Use Default Template option is checked (see Figure 2–16). This setting will use a specific part file as a seed file for the new object. Pro/ENGINEER's initial default template file for a new part includes a set of three default datum planes and the default coordinate system. The part file's default units is set at *Inch lbm Second.* This book assumes you will use this template file. The default template file for a part can be changed with the configuration file option *template_solidpart.*

STEP 4: **Enter a name for the new part file, then select OKAY on the New dialog box.**

STEP 5: **Select SET UP >> UNITS option on the main menu.**

STEP 6: **On the Units Manager dialog box, select *Inch Pound Second (IPS);* then select the SET icon.**

STEP 7: **On the Warning dialog box, select the CONVERT EXISTING NUMBERS option; then select OKAY.**

STEP 8: Close the Units Manager dialog box.

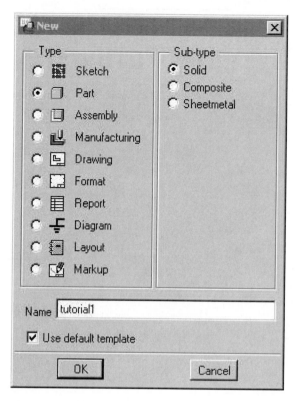

Figure 2-16   New dialog box

CREATING AN EXTRUDED PROTRUSION

Protrusion features are the most common positive space feature found in a Pro/ENGINEER solid model. Protrusions can consist of extruded, revolved, swept, or blended features. This segment of the tutorial will extrude a sketched section.

STEP 1:   **Select the FEATURE >> CREATE >> PROTRUSION command option.**

Protrusions are the most common type of geometry feature first created in a solid part.

STEP 2:   **Select the EXTRUDE >> SOLID >> DONE option.**

Extrude is one of the types of protrusions that can be created (see Figure 2–17). Other options available include Revolve, Sweep, Blend, Helical Sweep, and Swept/Blend. You will be creating a solid feature in this exercise. Solid is selected by default in Pro/ENGINEER.

STEP 3:   **Select ONE SIDE >> DONE as an Extrude option (Figure 2–17).**

This exercise will extrude the section in one direction from the sketching plane. Features may be extruded both directions by selecting Both Sides.

STEP 4:   **On the work screen, select Datum Plane FRONT as the sketching plane.**

This extruded section will be sketched on datum plane FRONT. You can select the label associated with this datum (FRONT), you can select any portion of the boundary of the datum, or you can select the datum on the model tree.

STEP 5:   **Select OKAY to accept the feature creation direction.**

After selecting the sketching plane, the red arrow shown on the work screen points from the datum plane in the positive direction. For one-sided extrusions, this is the direction of feature creation.

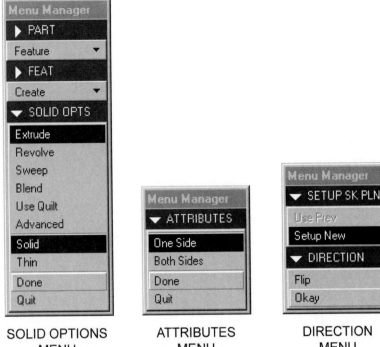

Figure 2-17   Menu options

STEP 6: **To orient the sketcher environment, select TOP on the Sketch View menu, then select datum plane TOP.**

You can orient your model in two different ways. The first method is by selecting Default. With default, Pro/ENGINEER determines the direction to orient the sketcher environment. The second method is to select a direction of orientation followed by a plane to face in the direction of orientation. If you select to use the latter method, select Top from the sketch view menu; then select datum plane TOP. This will orient datum plane TOP toward the top of the work screen.

SKETCHING THE SECTION

Extruded Protrusion features require a sketched section. Most extruded Protrusion features have a closed section. This segment of the tutorial will provide instruction on the sketching of the section for the extruded feature.

STEP 1: **Close the References dialog box.**

References are used by Pro/ENGINEER to pass design intent from parent features to child features. Pro/ENGINEER will automatically select a minimum number of references to allow for the sketching of a feature. You can use the References dialog box to add additional references or to delete references. In this example, datum planes RIGHT and TOP have been selected as references.

STEP 2: **Use the LINE icon to sketch the section shown in Figure 2–18.**

In Figure 2–18, the dimensioning scheme defining the size of the sketch has been purposely hidden. When sketching, you should not worry about the

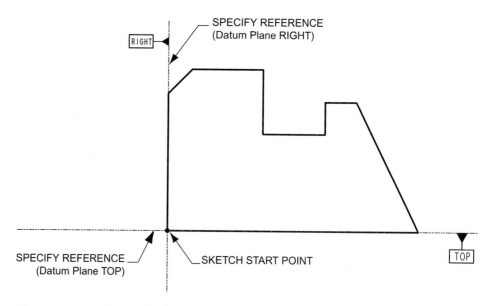

Figure 2-18   Sketched section of feature

size of your sketch. What is important is the sketching of geometry that matches the shape of your design intent. Start your sketch at the intersection of datum planes RIGHT and TOP, sketching in a counterclockwise direction. The left mouse button is used to pick entity locations, and the middle mouse button is used to cancel a command. Align the bottom edge of the feature with datum plane TOP, and align the vertical edge of the sketch with datum plane RIGHT.

STEP   3:   Use the ARC icon option to add a Tangent Arc to the sketch (see Figure 2–19).

STEP   4:   Delete the extra line shown in Figure 2–19 by first selecting the entity with the Pick icon, followed by selecting the Delete key on your keyboard.

STEP   5:   Use the CIRCULAR FILLET icon to add a Filleted Arc to the sketch.

Add an additional arc as shown in Figure 2–20. The Circular Fillet option will require you to select two nonparallel entities. A fillet will be created between the selected entities.

STEP   6:   Place Dimensions according to Design Intent.

Use the Dimension icon to match the dimensioning scheme shown in Figure 2–21. Placement of dimensions on a part should match design intent. With Intent Manager activated (Sketch >> Intent Manager), dimensions and constraints that fully define the section are provided automatically. Pro/ENGINEER does not know what dimensioning scheme will match design intent, though. Due to this, it is usually necessary to change some dimension placements.

MODELING POINT   If possible, a good rule of thumb to follow is to avoid modifying the section's dimension values until your dimension placement scheme matches design intent.

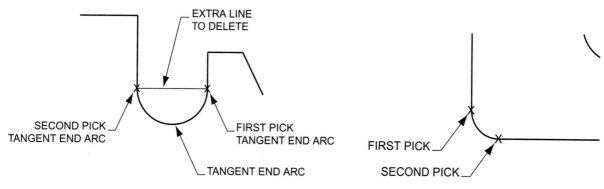

Figure 2-19 Tangent end arc   **Figure 2-20** Filleted arc

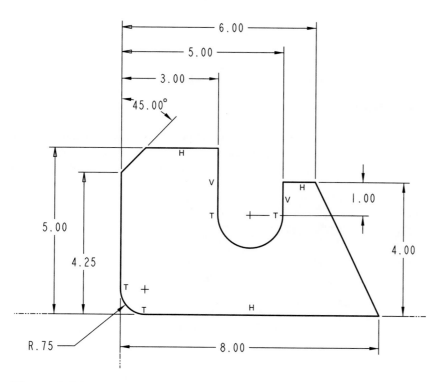

Figure 2-21 Dimensioning scheme

STEP 7: Using the Pick icon, drag a box around the entirety of the sketch making sure to include all dimensions (see Figure 2–22).

Your next task is to modify the sketch's dimension values. By preselecting all available dimensions, you will be able to simultaneously modify each dimension.

STEP 8: Select the Modify icon, then check the Lock Scale option (Figure 2–23).

The Lock Scale option will allow you to modify one dimension, with the remaining dimensions scaling the same factor.

STEP 9: Modify the first dimension value to equal 10, select ENTER on your keyboard, then select the CHECK icon on the dialog box.

At this point, it does not matter which dimension you modify to 10 units, though it is best to select the largest dimension. Your objective is to scale

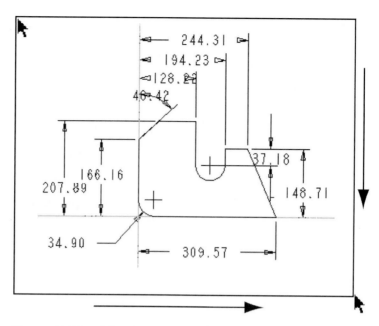

Figure 2-22   Select dimension using the pick icon

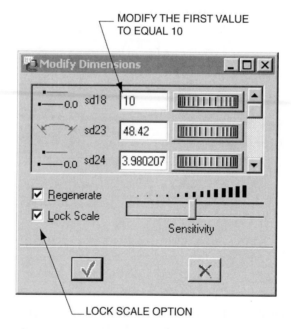

Figure 2-23   Dimension scale modification

the model down to a workable size. The next step of this tutorial will require you to modify each dimension individually.

Step 10:   Using the Pick icon, double select the arc's radius dimension value then modify its value to equal 0.750.

Double picking a dimension value will allow it to be modified through Pro/ENGINEER's textbox. Remember to select Enter on your keyboard after modifying the value. The sketch will be modified automatically. Often, it is best to modify smaller dimension values first. If your sketch has unexpected results, select the Undo icon on the toolbar.

STEP 11: Starting with the smallest dimension values first, modify the remaining dimensions to match Figure 2–21.

Use the Undo icon on the toolbar to undo unexpected results.

MODELING POINT   Sketched protrusion sections have to be completely enclosed and with no intersections. If a section has intersecting entities or open geometry, an open loop error will occur.

STEP 12:   ✔ Select the Continue icon to exit the sketching environment.

FINISHING THE FEATURE

The following steps define the process for finishing the creation of a feature.

STEP 1: Select BLIND >> DONE as the Depth option, then enter 2.00 as the Depth value.

STEP 2: Preview the feature.

On the Feature Definition dialog box (Figure 2–24), select Preview to observe your feature.

STEP 3: Dynamically view your part.

Shade your part by selecting the Shade icon from the toolbar menu. Dynamically view your part using the following options:

- Dynamic rotation   Control key and middle mouse button
- Dynamic pan   Control key and right mouse button
- Dynamic zoom   Control key and left mouse button

STEP 4: Select OKAY to finish the feature.

Selecting OKAY will create the feature. Observe the Model Tree to see the new feature (Figure 2–25).

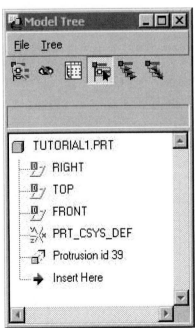

Figure 2-25   New feature added to model tree

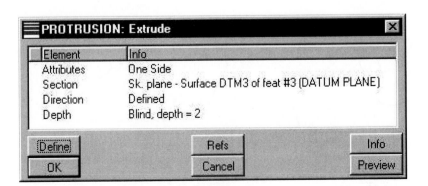

Figure 2-24   Feature Definition dialog box

STEP 5: Save your part file.

Use the File >> Save option to save your part file.

CREATING AN EXTRUDED CUT

This segment of the tutorial will create the extruded Cut feature shown in Figure 2–26. Additionally, the Define option will be used to modify the sketch.

STEP 1: Select FEATURE >> CREATE >> CUT.

The Cut option is a common way to remove material from existing features.

STEP 2: Select the EXTRUDE >> SOLID >> DONE option.

Extrude >> Solid is the default setting for Pro/ENGINEER. Select Done to continue with the command sequence.

STEP 3: Select ONE SIDE >> DONE as the extrude attribute for the feature.

The One Side option will extrude the section one direction from the sketching plane.

STEP 4: Pick the front of the part (Figure 2–27) as the sketching plane.

You will be sketching on the front of the part. Any planar surface can be used as a sketch surface.

> **MODELING POINT**   By selecting the first protruded feature as your sketching plane, this new Cut feature will become a child feature of the protruded feature. Any changes to a parent feature can affect its children.

STEP 5: Select OKAY to accept the direction of feature creation.

When using the Extrude Cut command, Pro/ENGINEER will attempt to determine the correct direction of feature creation. You should see a red arrow that points toward the interior of the part. This arrow points in the direction of extrusion. The Flip option can be used, if necessary, to change the direction.

STEP 6: Select TOP then pick the top of the part (Figure 2–27).

To orient the sketching environment, from the Sketch View menu, select Top, then pick the top of the part. When selecting a surface for orientation, the selected surface becomes a parent feature of the feature under

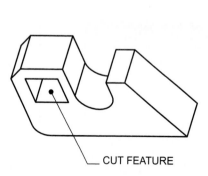

CUT FEATURE

Figure 2-26   Cut feature

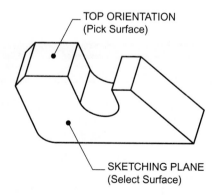

TOP ORIENTATION
(Pick Surface)

SKETCHING PLANE
(Select Surface)

Figure 2-27   Orienting the sketch

construction. Try to select an orienting surface that is already a parent feature.

STEP 7: **Utilizing the References dialog box, pick the two references shown in Figure 2–28; then close the dialog box.**

The design intent for this part requires this cut feature to be located from the top and left edges of the first part feature. As shown in Figure 2–28, select these two edges as references.

STEP 8: □ **Use the RECTANGLE icon to sketch the section.**

With the Rectangle option, you select diagonal corners of the rectangular feature.

STEP 9: |↔| **Apply the correct Dimensioning Scheme.**

Use the Dimension icon to apply the dimensioning scheme that matches the design intent. In this example the dimensioning scheme shown in Figure 2–29 matches the intent of the design.

STEP 10: **MODIFY dimension values.**

Use the Modify icon to change the dimension values to match Figure 2–29. Each dimension can be preselected with the Pick icon and the Shift key before selecting the Modify option. Another valid option would be to modify each dimension individually by double selecting it with the pick option.

STEP 11: **On the menu bar, use the SKETCH >> RELATIONS option to add a dimensional relationship.**

The design intent for the part requires that this cut feature remains square. Use the Add option on the Relations menu to add a relationship that will make the horizontal dimension equal to the vertical dimension. As shown in Figure 2–30, the dimension symbol for the vertical dimension is sd2, while the dimension symbol for the horizontal dimension is sd3. Your dimension symbols may be different. Select Add from the Relation menu. For the dimension symbols shown in Figure 2–30, enter the equation sd3 = sd2 in Pro/ENGINEER's textbox. Again, your dimension symbols may be different from those in the figure. Select Enter on the keyboard to exit the Add menu.

STEP 12: ✔ **Select the Continue icon to exit the sketcher environment.**

STEP 13: **Select OKAY to accept the material removal side.**

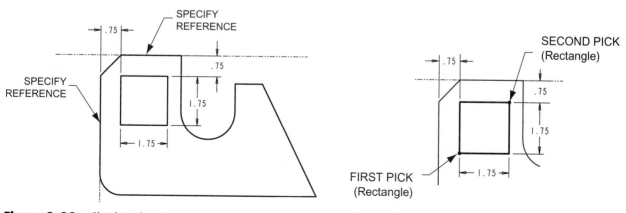

Figure 2-28   Sketching the cut   **Figure 2-29**   Sketching a rectangle

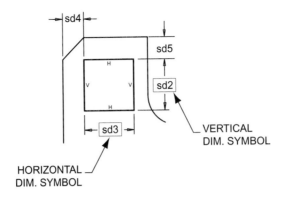

Figure 2-30   Dimension symbols

STEP 14:   Dynamically rotate the object.

Before selecting an extrude depth, dynamically rotate your object using the Control key and the middle mouse button. Observe the red arrow. This arrow points in the direction of extrusion.

STEP 15:   Select THRU ALL >> DONE as the depth option.

The Thru All depth option will create a cut whose depth always extrudes completely through any previously created features.

STEP 16:   On the Feature Definition dialog box, select the PREVIEW option (do not select OKAY).

Select Preview on the Feature Definition dialog box. Dynamically rotate your part to observe changes. At this time, **DO NOT SELECT OKAY** on the dialog box.

MODELING POINT   Previewing a feature is an important step during part modeling. Any conflicts that might exist between the new feature and existing features will normally be revealed during the preview process. If an error does occur, definitions associated with the feature can be redefined by using the Define option on the Feature definition dialog box.

REDEFINING THE FEATURE

In this section of the tutorial, you will modify the sketch of the Cut feature. If you inadvertently finished the feature by selecting OKAY in the previous step, you can access the Feature Definition dialog box by selecting the Feature >> Redefine option. Follow this by selecting the Cut feature on the model tree.

STEP 1:   Select the SECTION element on the Feature Definition dialog box.

Select the Section element on the dialog box (Figure 2–31). You will redefine the sketch of the Cut feature. The redefinition of the sketch is a suboption under Section.

STEP 2:   Select DEFINE on the Feature Definition dialog box.

With the Section option highlighted (Figure 2–31), select Define.

STEP 3:   Select SKETCH on the Section menu.

As shown on the menu, other options available in addition to Sketch include Scheme and Sketching Plane. The Scheme option is used to modify the

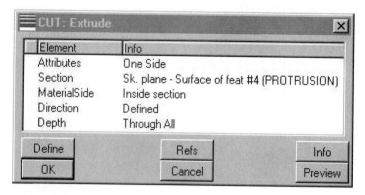

Figure 2-31   Cut Feature Definition dialog box

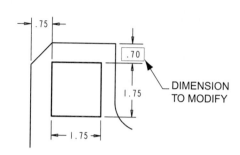

Figure 2-32   Modifying a dimension

dimensioning scheme of a sketch while the Sketching Plane option is used to change the planar surface upon which the section was sketched.

STEP 4:   Modify the dimension value shown in Figure 2–32.

As shown in Figure 2–32, use the Modify option to change the 0.75 dimension value to equal 0.70.

STEP 5:   Select the Continue icon to exit the sketching environment.

STEP 6:   Preview your part.

Select Preview on the Feature Definition dialog box. Dynamically rotate your model.

STEP 7:   Select OKAY on the dialog box to finish the feature.

STEP 8:   Save the part.

CREATING AN EXTRUDED CUT

This segment of the tutorial will create the Cut feature shown in Figure 2–33.

STEP 1:   Select FEATURE >> CREATE >> CUT.

STEP 2:   Select EXTRUDE >> SOLID >> DONE.

STEP 3:   Select BOTH SIDES >> DONE as an attribute of the feature.

This section of the tutorial will require you to extrude a cut in two directions from the sketching plane.

STEP 4:   Select the sketching plane upon which to sketch the Cut.

Select the planar surface shown in Figure 2–34 as the sketching plane.

STEP 5:   Observe the direction of Cut.

Notice the red arrow on your part protruding from the sketching plane. Since you will be cutting in both directions, this arrow represents the first direction of cut. This arrow is also important for the orientation of the sketcher environment. Within a sketcher environment, this arrow shows the direction that your sketch plane will be facing.

STEP 6:   Select OKAY to accept the first direction of cut.

STEP 7:   Select LEFT; then pick the plane shown in Figure 2–34 to orient your sketching environment.

By selecting **Left** and by selecting the front of the part, you will be orienting the front of the part toward the left of the sketcher environment.

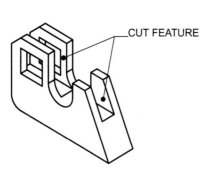

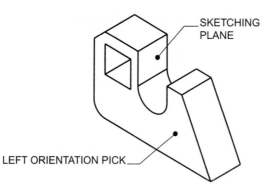

Figure 2-33   Cut feature

Figure 2-34   Sketching plane and left orientation

SKETCHING THE SECTION

This segment of the tutorial will sketch the Cut section shown in Figure 2–35.

STEP   1:   **On Pro/ENGINEER's toolbar, turn off the display or datum planes.**

STEP   2:   **Use the References dialog box to specify the two references shown in Figure 2–35.**

If necessary, you can also use the References dialog box to delete unwanted references.

STEP   3:   **Use the LINE icon options to sketch the three lines representing the section.**

Start the first line by using Line option. Make sure that the start point and endpoint of the sketch are aligned with existing geometry.

STEP   4:   **Dimension geometry according to design intent.**

Figure 2–35 portrays design intent for the feature. Use the **Dimension** icon to dimension the feature.

STEP   5:   **MODIFY dimension values.**

Use the Modify option to modify your dimension values, or double pick each dimension with the Pick icon.

STEP   6:   **Select the Continue icon to exit the sketcher environment.**

If you do not have a fully defined section, Pro/ENGINEER will give you an error message stating this fact. If this occurs, try to realign your elements or to add dimensions that will fully define the sketch.

STEP   7:   **Select OKAY on the Material Direction menu.**

Pro/ENGINEER tries to determine the side of the section from which to remove material. As shown in Figure 2–36, an arrow points in the direction of material removal. If this is not the correct side, you can Flip the arrow.

STEP   8:   **Dynamically Rotate your model.**

Dynamically rotate your model to observe the direction of cut. Notice the direction that the arrow points. Since you are performing a Both Sides extrusion, this will be your first cut direction.

STEP   9:   **Select THRU ALL >> DONE as the first Cut depth.**

STEP 10:   **Select THRU ALL >> DONE as the second Cut depth.**

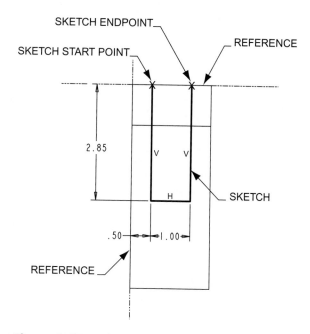

Figure 2-35   Sketching the cut

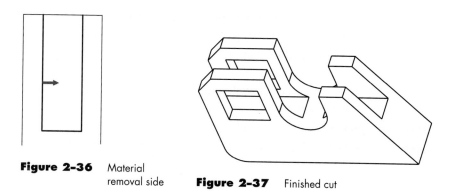

Figure 2-36   Material removal side

Figure 2-37   Finished cut

STEP 11:   Preview your part.

Your part should look as shown in Figure 2–37.

MODELING POINT   If you missed the step to perform a Both Sides extrusion, it is not too late to change this attribute. You can select the Attributes element in the Feature Definition dialog box and use the Define option to change from One Side to Both Sides.

STEP 12:   Select OKAY to finish the part.

STEP 13:   SAVE your part.

CREATING AN EXTRUDED PROTRUSION

This segment of the tutorial will create the extruded Protrusion feature shown in Figure 2–38. Within this tutorial, the Use Edge option will be introduced. The Use Edge option turns existing part features and edges into sketch geometry.

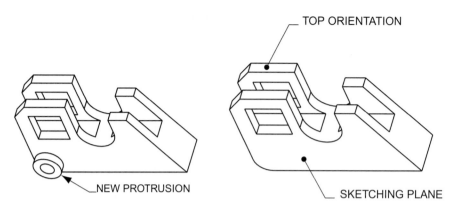

Figure 2-38   Extruded feature   **Figure 2-39**   Sketching plane and orientation

STEP   1:   Select the FEATURE >> CREATE >> PROTRUSION option.

This section of the tutorial will create a Protrusion on the front surface of the part.

STEP   2:   Select the EXTRUDE >> SOLID >> DONE option.

STEP   3:   Select ONE SIDE >> DONE as an attribute of the feature.

Your protrusions will extrude in one direction from the sketch plane.

STEP   4:   Select the front of the part as the sketching plane (Figure 2–39).

STEP   5:   Observe the direction of Protrusion.

The arrow shown on the work screen points in the direction that your sketch plane will be facing while sketching. It also represents the direction of feature creation.

STEP   6:   Select OKAY to accept the extrude direction.

STEP   7:   Select TOP; then pick the top of the part (Figure 2–39) to orient the sketching environment.

The Top option will orient the selected surface toward the top of the sketcher environment. After selecting the orienting surface, Pro/ENGINEER will launch the sketching environment.

STEP   8:   On the toolbar, turn off the display of datum planes.

STEP   9:   Utilizing the USE EDGE icon, select the edge shown in Figure 2–40. (Make sure you only select the edge once.)

During this portion of the tutorial, you will select an existing edge to project as an entity in this feature's section. The Use Edge option will turn existing feature edges into entities of the current sketch. When selecting existing feature edges, the feature from which the edge is obtained becomes a parent feature of the feature under construction.

Only select the edge once. Since the newly formed sketch entity will lie coincident with its parent edge, it may not be easily identifiable. Selecting a second time will create a second entity. This will cause an error when you attempt to exit the sketcher environment.

STEP 10:   Use the ARC icon to create the tangent end arc shown in Figure 2–40.

You will create an arc that joins the arc created in the previous step. In combination with the arc from the previous step, this newly created arc will

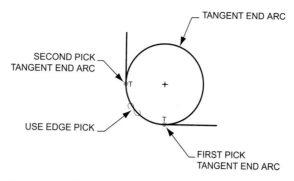

Figure 2-40   Tangent end arc

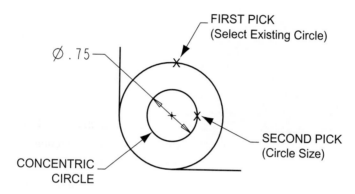

Figure 2-41   Concentric circle

form a circle. As shown in Figure 2–40, select at the first tangent point, then the second tangent point. When moving the cursor from Point 1 to Point 2, ensure that the arc is created toward the inside of the part.

Step 11:   Use the CONCENTRIC circle icon to create the circle shown in Figure 2–41.

The Concentric Circle icon can be found under the normal Circle icon on the sketch toolbar. Perform the two picks shown in the figure, then select the middle mouse button to end the Concentric Circle creation process.

Step 12:   MODIFY the circle's diameter dimension value.

Modify the circle's dimension value to equal .75.

Step 13:   Select the Continue icon to exit the sketcher environment.

After selecting continue, if you get the error message *Cannot have mixture of open and closed sections,* you probably picked the edge more than once while executing the Use Edge option. If necessary, use the Pick option and the delete key to delete any extra entities.

Step 14:   Observe the direction of protrusion.

Notice the two concentric circles formed on the inside of the sketched circle. These circles represent the pointing end of the extrusion direction arrow. Dynamically rotate the part to get a better view of the extrusion direction.

Step 15:   Select BLIND >> DONE as the extrude depth.

The Blind option will allow you to enter an extrude depth.

Figure 2-42　Preview of model

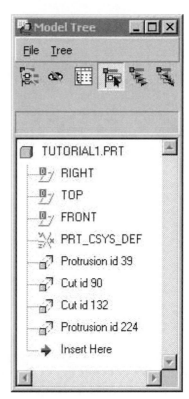

Figure 2-43　Model tree

STEP 16: **Enter .500 as the Blind extrude depth.**

In Pro/ENGINEER's Textbox, enter .500 as the Blind extrude depth.

STEP 17: **Preview your model.**

Select the Preview option on the Feature Definition dialog box.

STEP 18: **Shade your model and dynamically rotate to observe the protrusion.**

Your model should appear as shown in Figure 2–42.

STEP 19: **Select OKAY on the Feature Definition dialog box.**

STEP 20: **Observe the Model Tree.**

The Model Tree for the part is shown in Figure 2–43. This part contains, in order of creation, a Protrusion, a Cut, a second Cut, and a second Protrusion. This is reflected on the Model Tree.

STEP 21: **SAVE your part.**

DIMENSION MODIFICATION

This segment of the tutorial will modify the dimensions used to define the cut feature. Covered will be options for modifying a dimension's value and tolerance format.

STEP 1: **Select MODIFY >> VALUE on Pro/ENGINEER's Menu Manager.**

STEP 2: **On the Model Tree pick the second Cut feature (Figure 2–43).**

This step of the exercise requires you to select the second cut feature. You may select this feature directly on the work screen, or you may select the

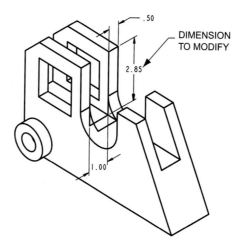

Figure 2-44 Dimension modification

feature from the model tree. When you select the feature, parametric dimensions used to construct the feature will appear (Figure 2–44).

MODELING POINT Often, it is difficult to select a feature directly on the work screen. When it is necessary to select a feature or element directly on the work screen, consider using the Query Sel option from the Get Select menu. Query select will give you an option to toggle through available features.

STEP   3: **Pick the 2.85 dimension and modify its value to equal 2.90.**

After picking the dimension, Pro/ENGINEER requires you to enter the new dimension value in the textbox.

STEP   4: **REGENERATE the part.**

After modifying a dimension value, Pro/ENGINEER requires a regeneration of the part. Select Regenerate from the Part menu.

STEP   5: **On the menu bar, select the UTILITIES >> ENVIRONMENT option.**

STEP   6: **On the Environment dialog box, check the DIMENSION TOLERANCES setting; then select OKAY.**

The Environment dialog box is located under the Utilities menu bar. Check the Dimension Tolerance setting. This will display dimensions with tolerances.

STEP   7: **Select MODIFY >> DIMCOSMETICS from Pro/ENGINEER's Menu Manager.**

STEP   8: **Select the FORMAT >> PLUS-MINUS format type.**

STEP   9: **On the Model Tree, select the last cut feature.**

STEP 10: **To set a Plus-Minus Tolerance Mode, select each dimension defining the Cut feature (Figure 2–45).**

Selected dimensions will be displayed in a Plus-Minus tolerance format. Be sure to only select the text. While selecting dimension values, you might inadvertently pick a geometric feature on the work screen. This will cause dimensions associated with this feature to be displayed also.

STEP 11: **Select DONE SEL on the Get Select menu.**

Done Sel will allow you to finish selecting dimensions to modify. The middle mouse button serves the same purpose.

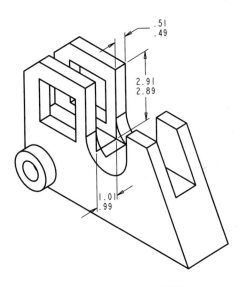

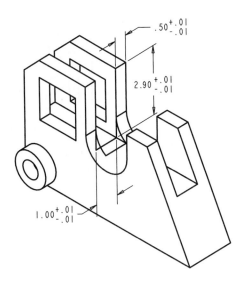

TOLERANCE VALUES
IN LIMIT FORMAT

TOLERANCE VALUES
IN PLUS-MINUS FORMAT

Figure 2-45   Dimension tolerance values

Figure 2-46   Dimension Properties dialog box

STEP 12: On the Modify menu, select the DIMENSION option.

STEP 13: On the Model Tree or work screen, select the second Cut feature.

STEP 14: Select the three dimensions defining the Cut feature; then select the DONE SEL option on the Get Select menu.

When selecting dimensions, you must select the nominal value of each displayed dimension since tolerances are displayed.

STEP 15: On the Dimension Properties dialog box, enter 3 as the number of decimal places.

As shown in Figure 2–46, under the Dim Format option, enter 3 as the number of decimal places.

STEP 16:   Enter Upper and Lower Tolerance values.

As shown in Figure 2–46, enter 0.005 as the Upper Tolerance value and 0.000 as the Lower Tolerance value. Notice, on the dialog box, how the Tolerance Mode is set to Plus-Minus.

> **MODELING POINT**   Other options are available on the Modify Dimension dialog box. The DimText tab is used for entering additional text, notes, and symbols around an existing dimension value. A Sym Pallette (Symbol Pallette) option is used to select predefined graphical symbols that can be used under the Dim Text tab.

STEP 17:   Select the OKAY option on the dialog box.

STEP 18:   Save your model.

REDEFINING A FEATURE'S DEPTH

This segment of the tutorial will redefine the extrusion depth of the last Protrusion feature. You will change the depth from a value of 0.500 to a value of 1.00. The Redefine command is one of Pro/ENGINEER's most useful and powerful options. It is used to redefine attributes and parameters associated with features. Feature parameters such as Feature Creation Direction, Material Removal Side, and Depth value can be modified.

STEP 1:   Select the FEATURE >> REDEFINE command.

STEP 2:   On the Model Tree, select the last Protrusion feature (Figure 2–47).

As shown on the model tree, this tutorial has required the creation of two protrusion features and two cut features. This section of the tutorial will require you to modify the last protrusion feature.

STEP 3:   On the Protrusion Feature Definition dialog box, select the DEPTH definition (Figure 2–47).

STEP 4:   Select the DEFINE option on the dialog box.

After selecting the Define option, Pro/ENGINEER will display the depth option menu. This will allow you to redefine the feature's depth parameter.

STEP 5:   Select BLIND >> DONE on the depth option menu.

Notice on the Feature Definition dialog box that the attribute associated with the Depth option is now showing a value equal to *Changing*.

STEP 6:   Enter 1.00 as the new Depth value.

STEP 7:   Select PREVIEW on the Feature Definition dialog box.

STEP 8:   Select OKAY to accept the changes.

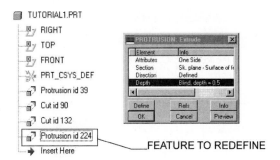

Figure 2-47   Redefining a feature

REDEFINING A FEATURE'S SECTION

This section of the tutorial will use the Redefine option to add filleted corners to the first Cut feature.

STEP   1:   **On the model tree, select the first cut feature with your right mouse button; then select the REDEFINE command.**

Within this segment, you will access the Redefine command through the model tree. Optionally, you could select the Redefine command on the menu, followed by selecting the protrusion feature.

STEP   2:   **As shown in Figure 2–48, select the SECTION definition on the Feature Definition dialog box.**

STEP   3:   **Select the DEFINE option on the dialog box.**

STEP   4:   **Select SKETCH on the Section menu.**

The Section menu gives you the option of modifying the sketch of the feature, the dimensioning scheme of the sketch, or the sketch plane for the section. After selecting Sketch, Pro/ENGINEER will take you into the sketcher environment that defined the original feature.

STEP   5:   ⌁ **Using the CIRCULAR FILLET icon, construct the filleted arcs shown in Figure 2–49.**

The Circular Fillet option requires you to pick the two entities bordering the arc. The first pick will define the initial radius of the fillet.

STEP   6:   ⇨ **Modify the radius dimensions of each fillet to equal 0.500.**

When selecting a dimension value to modify, Pro/ENGINEER requires that you select the text of the dimension. Your section should appear as shown in Figure 2–50.

STEP   7:   **On the Menu bar, select the SKETCH >> RELATION option.**

STEP   8:   **On the Relations menu, select the EDIT REL option to adjust dimension symbols.**

The original sketch was created with the Rectangle sketch option. The default scheme dimensions the length of each line of the rectangle. A previous step of this tutorial required the establishment of a dimensional relationship between the two defining dimensions. The Circular Fillet option will create new dimensions with new dimension symbols. Use the Edit Rel option to adjust the symbols defining the dimensional relationship.

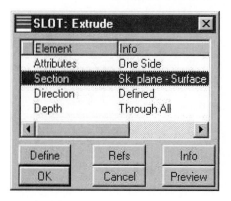

Figure 2-48   Cut Feature Definition dialog box

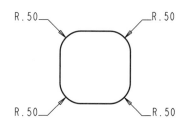

Figure 2-49   New section

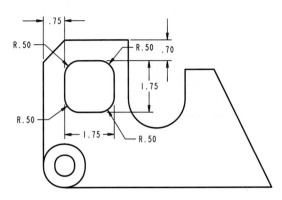

Figure 2-50   Final section

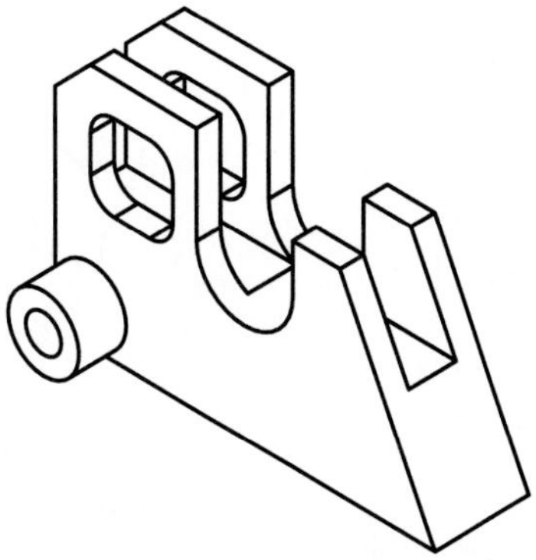

Figure 2-51   Completed part

Step   9:   Select the Continue icon to exit the sketching environment.

Step 10:   Select PREVIEW on the Feature Definition dialog box.

You final part should look as shown in Figure 2–51.

Step 11:   SHADE and Dynamically rotate the part.

Step 12:   Select OKAY on the Feature Definition dialog box.

Step 13:   SAVE your part.

Step 14:   Use the FILE >> DELETE >> OLD VERSIONS option to purge old versions of the part.

Every save of the part creates a new version. Use the File >> Delete >> Old Versions option to delete every version except for the last saved.

PROBLEMS

1.   Use Pro/ENGINEER to model the parts shown in Figures 2–52 through 2–56.

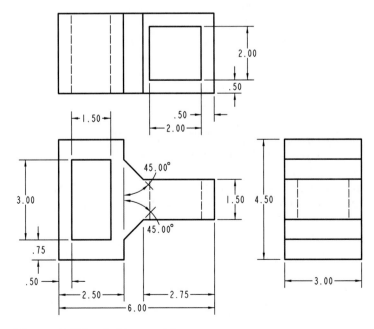

Figure 2-52   Problem one (a)

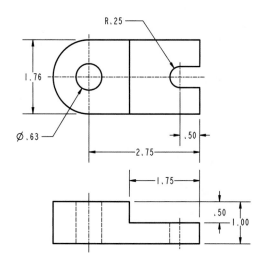

Figure 2-53   Problem one (b)

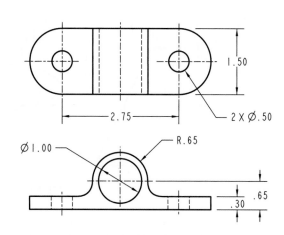

Figure 2-54   Problem one (c)

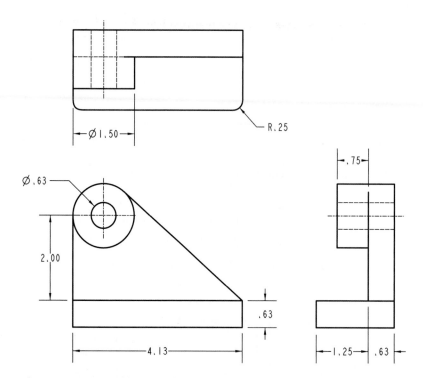

Figure 2-55   Problem one (d)

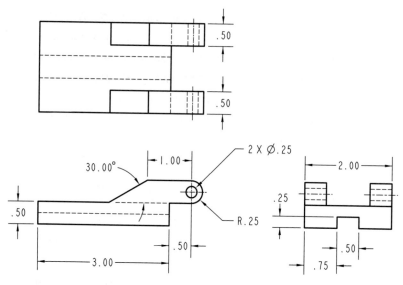

Figure 2-56   Problem one (e)

2. Use Pro/ENGINEER to model the part shown in Figure 2–57. Use millimeter-newton-seconds as the units for the part.

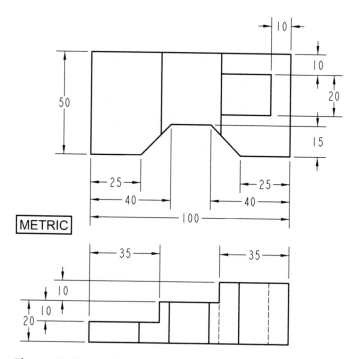

Figure 2-57   Problem two

QUESTIONS AND DISCUSSION

1.   Describe each of the feature creation options found under the Protrusion and Cut commands.

2.   What is the difference between a solid feature and a thin feature?

3.   How does a Both Sides extrusion differ from a One Side extrusion?

4.   Compare the Thru Until depth option with the UpToSurface depth option.

5.   Describe the methods available within Part mode to modify a dimension's value.

6.   What option within Pro/ENGINEER is available for changing the depth definition for an extruded feature?

3

FEATURE CONSTRUCTION TOOLS

Introduction

Pro/ENGINEER provides many feature construction tools that do not rely upon a sketcher environment. Examples include holes, rounds, and chamfers. This chapter will cover the basics of these and other common feature construction commands. Upon finishing this chapter, you will be able to

- Construct a straight-linear hole.
- Construct a straight-coaxial hole.
- Construct a standard hole.
- Create fillets and rounds on a part.
- Create a chamfer on a part.
- Tweak features using the draft option.
- Create a part shell.
- Create a cosmetic thread.
- Create a linear pattern.

DEFINITIONS

Draft surface   A part surface angled to allow for easy removal of the part from a mold or cavity.

Cosmetic features   A part feature that allows for cosmetic details that do not require complicated regenerations. An example of a cosmetic feature would be a cosmetic thread or a company logo sketched on a surface.

Sketched hole   A hole, such as a counterbored or countersunk hole, that has a sketched profile.

HOLE FEATURES

The Hole command is used to create either straight holes, holes with varying profiles, or holes defined by standard fastener tables. A hole is considered a negative space feature. Straight holes can be created with the Cut command also. Why use the Hole command instead of Cut? The Hole command has a predefined section that does not require sketching. Additionally, the Hole command's algorithms are not as complicated as the Cut command's, thus requiring less computer power to regenerate.

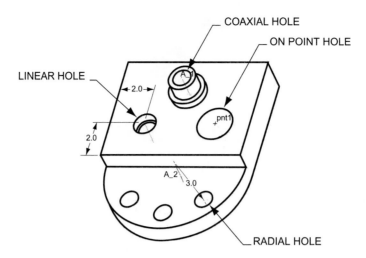

Figure 3-1   Hole placement options

HOLE PLACEMENT OPTIONS

Four hole placement options are available (Figure 3–1). A description of each follows:

LINEAR

The Linear option locates a hole from two feature edges. These edges may be part surfaces or datum planes. For each location edge, the user must pick an edge and then enter a dimension value.

RADIAL

The Radial option locates a hole at a distance from an axis and at an angle from a reference plane. This option is used often for radial patterns such as might be found on a bolt-circle pattern. This option is covered in detail in the chapter on revolved features.

COAXIAL

The Coaxial option locates the center of a hole coincident with an existing axis. The user must provide the axis, placement surface, and hole diameter.

ON POINT

The On Point option locates the center of a hole on a datum point. The user must provide the point, placement surface, and hole diameter.

HOLE TYPES

Pro/ENGINEER provides three options for defining the profile of a hole: Straight, Sketched, and Standard. The Straight option produces a hole that has a constant diameter throughout the length of the hole. The Sketched option requires the user to sketch the profile of the hole within a sketcher environment. Figure 3–2 shows an example of a sketched profile of a counterbored hole. The Sketched Hole option is covered in Chapter 4.

A standard hole is defined through a fastener table. An example would be a 1.00 unified national course fastener. It has 8 threads per inch with a tap drill of 7/8 inch. As shown in Figure 3–3, a Standard hole is specified along with a tapped screw size equal to *1-8.* This will produce an interior thread minor diameter equal to 0.875 (or 7/8 inch). Pro/ENGINEER derives this information from three default text files: *ISO.hol, UNC.hol.* and *UNF.hol.* Additional user-defined hole charts can be created and specified with the configuration file option *hole_parameter_file_path.*

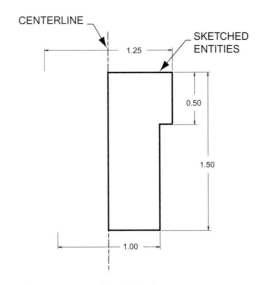

Figure 3-2   Sketched hole

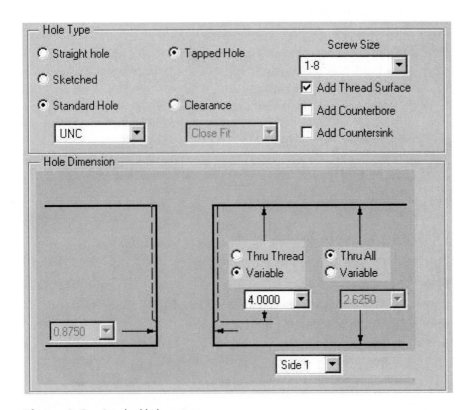

Figure 3-3   Standard hole options

The following options are available for standard holes.

- **Tapped Hole** The Tapped Hole option is selected when a fastener's threads are engaged within the hole. Setting this option will provide hole parameters for an internal thread (e.g., minor diameter and thread length).

- **Clearance** The Clearance option is selected when a fastener's threads are not engaged within the hole. Setting this option provides a clearance fit for the fastener.

- **Add Thread Surface** This option is only available when the Tapped Hole option is selected. It provides a cosmetic representation of the thread. A Clearance hole has Thru All as its thread depth.
- **Add Counterbore** As the name implies, the Add Counterbore option creates a counterbored hole. The user must enter the counterbore's diameter and depth.
- **Add Countersink** Similar to the Add Counterbore option, the Add Countersink option creates a countersunk hole. The user must enter parameters for the hole.

HOLE DEPTH OPTIONS

Similar to the Extrude option, straight holes have several depth options available. As with extruded features, the selection of a hole's depth option is important for incorporating design intent. As an example, it is common for holes to cut completely through a part. If this is the intent of a design then the Thru All option should be used. The following are available depth options.

- **Variable** The Variable option cuts a hole to a user-defined depth. The constructed hole has a flat bottom.
- **Thru Next** This option cuts the hole to the next part surface.
- **Thru All** This option cuts a hole completely through a part.
- **Thru Until** This option cuts a hole to a user-selected surface. The constructed hole has a flat bottom.
- **To Reference** This option cuts a hole to a user-selected point, vertex, curve, or surface. The reference has to exist before executing this option. The hole will have a flat bottom after construction.
- **Symmetric** This option creates a two-sided hole with equal depths on both sides of the placement plane. This option is only selectable under the Depth Two parameter.

CREATING A STRAIGHT LINEAR HOLE

Perform the following steps to create a Straight Linear Hole:

STEP   1:   Select the FEATURE >> CREATE >> HOLE command.

STEP   2:   On the Hole dialog box, select STRAIGHT as the Hole Type.

STEP   3:   Enter a Diameter value for the hole.

STEP   4:   Select a Depth One parameter (e.g., Thru All, Variable, etc.), then perform the operation appropriate to the option.

INSTRUCTIONAL NOTE   If you select a depth parameter such as Variable, Thru Until, etc., you will have to perform an additional step. As an example, if you select the To Reference option, you will have to pick a point, curve, or surface to cut the hole to.

STEP   5:   If required, select a Depth Two parameter, then perform the operation appropriate to the option.

The Depth Two parameter creates a Both Sides extruded hole.

STEP   6:   Select a Primary Reference for the placement of the hole (Figure 3–4).

The Primary Reference parameter defines the hole's placement plane.

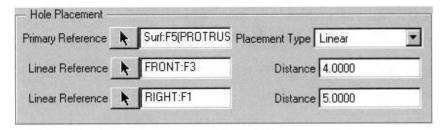

Figure 3-4   Linear hole placement references

STEP 7: On the work screen, select a reference edge or surface for locating the hole.

The Linear option requires the selection of two edges or planes to locate the hole. Pick a part edge or surface to position the hole in the first direction.

STEP 8: On the Hole dialog box, enter a distance to locate the hole from the first linear reference.

STEP 9: On the work screen, select a reference edge or surface for locating the hole.

Enter a value for the distance that the hole will be located from the selected edge.

STEP 10: On the Hole dialog box, enter a distance to locate the hole from the second linear reference.

STEP 11: Select the Build Feature checkmark.

CREATING A STRAIGHT COAXIAL HOLE

The Coaxial option locates the center of a hole coincident with an existing axis. Perform the following steps to create a straight-coaxial hole.

STEP 1: Select FEATURE >> CREATE >> HOLE.

STEP 2: Select STRAIGHT hole type.

A Coaxial hole can be either a Straight Hole, a Sketched Hole, or a Standard Hole.

STEP 3: Enter a Diameter value for the hole.

STEP 4: Select a Depth One parameter (e.g., Thru All, Variable, etc.), then perform the operation appropriate to the option.

> **INSTRUCTIONAL NOTE**   If you select a depth parameter such as Variable, Thru Until, and so forth, you will have to perform an additional step. As an example, if you select the To Reference option, you will have to pick a point, curve, or surface to cut the hole to.

STEP 5: If required, select a Depth Two parameter, then perform the operation appropriate to the option.

The Depth Two parameter creates a Both Sides extruded hole.

STEP 6: On the work screen, select the axis along which to place the hole's centerline.

The axis selection for a coaxial hole is the primary reference.

STEP 7: Select the Hole's placement plane.

STEP 8: Select the Build Feature checkmark.

ROUNDS

Pro/ENGINEER utilizes the Round command to create both fillets and rounds. Despite its apparent simplicity, the Round command can be one of Pro/ENGINEER's most difficult tools to master. Rounds constructed on complex features often result in failures. The following modeling techniques should be used to help avoid round conflicts.

- Create rounds toward the end of the modeling process.
- Create smaller radii rounds before larger.
- Avoid using round geometry as references for the creation of features.
- If a surface is to be drafted, draft the surface first, then create any necessary rounds.

ROUND RADII OPTIONS

Several different Round options are available within Pro/ENGINEER (Figure 3–5).

- **Constant** The Constant option creates a round with a constant radius.
- **Variable** The Variable option creates a round with a variable radius. Radii values are defined from the end of chained segments.
- **Thru Curve** The Thru Curve option defines a round's radius based on a selected curve.
- **Full Round** The Full Round option creates a round in place of a selected surface.

ROUND REFERENCE OPTIONS

Rounds are normally created on the edge of a feature. Pro/ENGINEER provides the Edge Chain option to perform this task. Other options are available to allow for flexibility in the round creation process.

- **Edge Chain** This option allows for the selection of edges to place rounds. Sub-options exist for selecting edges one at a time or for selecting tangent edges.
- **Surf-Surf** This option allows for the selection of two surfaces to place a round. The round will be formed between the two surfaces.

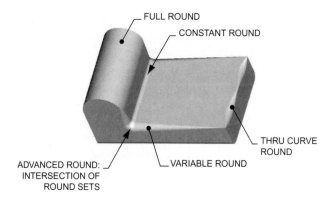

Figure 3-5   Round options

- **Edge-Surf** This option places a round between a selected surface and a selected edge.
- **Edge Pair** This option is similar to the Full Round radius option. With this option, the surface between two selected edges will be replaced with a round.

CREATING A SIMPLE ROUND

Perform the following steps to create a simple Round while utilizing the Edge Chain option.

STEP 1: **Select FEATURE >> CREATE >> ROUND.**

STEP 2: **Select SIMPLE >> DONE on the Round Type menu.**

An advanced round type is also available. The Advanced option allows for more complex rounds. As an example, advanced rounds often have varying cross sections and multiple round sets.

STEP 3: **Select the CONSTANT >> EDGE CHAIN >> DONE options on the Round Set Attribute menu.**

Round Radius options available include Variable, Thru Curve, and Full Round. Other Round Reference options include Surf-Surf, Edge-Surf, and Edge Pair.

STEP 4: **Select ONE-BY-ONE on the Chain menu.**

The following chain options exist:

- **One By One** Individual edges are selected.
- **Tangent Chain** Edges that lie tangent are selected (default selection).
- **Surf Chain** Edges are defined by selected surfaces.
- **Unselect** This option allows for the unselection of a reference.

STEP 5: **Select feature edges to round.**

Due to the **One-By-One** option selection, each edge to round must be selected.

STEP 6: **Select DONE on the Chain menu.**

When you are through selecting edges to round, select Done.

STEP 7: **Enter a Radius value for the edges.**

STEP 8: **Preview the feature on the Round Feature definition dialog box.**

Previewing a feature, especially rounds, is an important step. Most failed features will be revealed with the Preview option. Additional options are available for resolving any conflicts.

STEP 9: **Select OKAY on the Round Feature definition dialog box.**

CHAMFER

Pro/ENGINEER provides a construction option for creating edge and corner chamfers (Figure 3–6). An Edge Chamfer creates a beveled surface along a selected edge. The Corner Chamfer option creates a beveled surface at the intersection of three edges. Perform the following steps to create an Edge Chamfer:

STEP 1: **Select FEATURE >> CREATE >> CHAMFER.**

The Chamfer command creates a beveled surface at a selected solid edge. Creating a chamfer does not require entering a sketching environment.

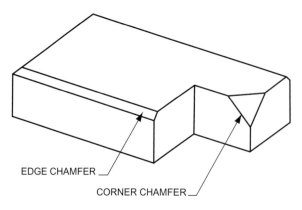

EDGE CHAMFER

CORNER CHAMFER

Figure 3-6   Chamfer types

STEP 2:   **Select EDGE as the chamfer method.**

You have the option of selecting either Edge or Corner (Figure 3–6).

STEP 3:   **Select an edge chamfer dimensioning scheme.**

The following dimensioning schemes are available.

- **45 × d** This option creates a chamfer with a 45-degree angle and with a user-specified distance

- **d × d** This option creates a chamfer at a user-specified distance from an edge.

- **d1 × d2** This option creates a chamfer at two specified distances from an edge.

- **Ang × d** This option creates a chamfer at a user-specified distance from an edge and at a user-defined angle.

STEP 4:   **In Pro/ENGINEER's textbox, enter appropriate dimension values.**

STEP 5:   **Select Edges to chamfer.**

A single edge or multiple edges can be selected. Use the Query Sel option for edge selection if necessary.

STEP 6:   **Select DONE SEL on the Get Select menu.**

Select Done Select after selecting edges to chamfer.

STEP 7:   **Select DONE REFS on the Feature Refs menu.**

The Feature Refs menu allows for the selection of additional edges to chamfer.

STEP 8:   **Select PREVIEW on the Feature Definition dialog box.**

MODELING POINT   Chamfer parameters such as selected edges and dimensioning scheme can be modified within the Chamfer Feature definition dialog box. Select the parameter to modify then select the Define button.

STEP 9:   **Select OKAY on the Feature Definition dialog box.**

DRAFT

Cast and molded parts often require a **drafted surface** for ease of removal from the mold. Under the Tweak menu option (Feature >> Create >> Tweak), Pro/ENGINEER provides a command for creating drafted surfaces. The maximum angle that may be created is plus-or-minus 30 degrees.

MODELING POINT   The Tweak menu provides a variety of options for modifying an existing part surface. Draft is just one option available. Other options include Offset for offsetting a surface and Radius Dome for creating a dome from a selected surface.

NEUTRAL PLANES AND CURVES

Selected surfaces are drafted by pivoting around a neutral plane or curve. Planes may be surface planes or datum planes, while curves may be datum curves or edges. Additionally, a surface can be split at the neutral plane or curve.

NEUTRAL PLANE DRAFTS

With a Neutral Plane Draft, picked surfaces are pivoted around a selected neutral plane. The plane can be a part surface or a datum. Three split options are available.

NO SPLIT

The No Split option creates a draft without a split in the drafted surface (Figure 3–7). The user selects the neutral plane and the draft surface, then enters a draft angle. The angle can be positive or negative.

SPLIT AT PLANE

The Split at Plane option creates a draft with the drafted surface split at the neutral plane (Figure 3–7). The portion of the surface selected for drafting is the surface that the draft angle is applied to. The user selects the neutral plane and the draft surface, then enters a draft angle.

SPLIT AT SKETCH

The Split at Sketch option creates at drafted surface out of a user-sketched section (Figure 3–7). The sketched section is pivoted around the neutral curve. The user selects the neutral plane and a surface to sketch upon. A sketcher

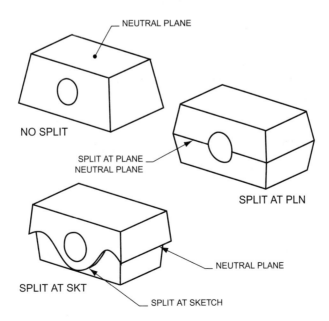

Figure 3-7   Neutral plane drafts

environment provides the user with the requirement of sketching the surface to be drafted.

CREATING A NEUTRAL PLANE, NO SPLIT DRAFT

Perform the following steps to create a neutral plane draft.

STEP 1: Select the FEATURE >> CREATE >> TWEAK menu option.

STEP 2: Select the DRAFT command on the Tweak menu.

STEP 3: Select NEUTRAL PLN >> DONE on the Draft Options menu.

Draft options available include Neutral Plane for pivoting around a plane or surface and Neutral Curve for pivoting around a datum curve or edge.

STEP 4: Select TWEAK >> NO SPLIT >> CONSTANT >> DONE on the Draft Attributes menu.

Using the Intersect option in place of Tweak would allow the drafted surfaces to overhang the edge of an adjacent surface. No Split will create a draft on the complete side of a selected surface. The Split at Plane option will create a split surface at the neutral plane, while the Split at Sketch option will create a draft out of a user-sketched surface.

STEP 5: Select surfaces to draft.

The default option is to select individual surfaces.

STEP 6: Select DONE on the Surface Select menu.

STEP 7: Select a NEUTRAL PLANE (Figure 3–8).

The Neutral Plane is the part surface that will remain intact during the drafting procedure. Since the drafted surfaces will pivot from the neutral plane, this is a critical selection. The Make Datum option exists for creating an on-the-fly datum plane.

STEP 8: Select a reference perpendicular plane (Figure 3–8).

The reference plane must lie perpendicular to the draft surfaces. Often, this reference plane is the same as the neutral plane.

STEP 9: Enter a draft angle.

As shown in Figure 3–9, Pro/ENGINEER will graphically display the positive direction of rotation. A negative value can be entered.

STEP 10: Preview the draft on the Draft Feature Definition dialog box.

STEP 11: Select OKAY on the dialog box to complete the draft.

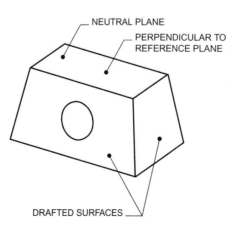

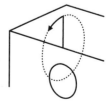

Figure 3-8   Neutral plane no split draft   **Figure 3-9**   Draft angle direction

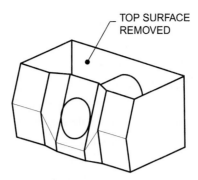

Figure 3-10   Shelled feature

SHELLED PARTS

The Shell command removes a selected surface from a part and hollows the part with a user-defined wall thickness. The wall is created around the outside surfaces of the part, with any features created before the shell feature being included (Figure 3–10). Perform the following steps to create a shelled part.

STEP 1:   **Select FEATURE >> CREATE >> SHELL.**

STEP 2:   **Select surfaces to remove.**

> The Shell command functions by removing selected surfaces from the part. Select surfaces to remove.

STEP 3:   **Select DONE SEL on the Get Select menu.**

> The Done Sel option ends the selection of surfaces to be removed by the Shell command.

STEP 4:   **Select DONE REFS from the Feature Refs menu.**

> The Feature References menu allows for the addition or removal of selected shell surfaces.

STEP 5:   **Enter a shell wall thickness.**

STEP 6:   **Select PREVIEW on the feature definition dialog box.**

STEP 7:   **Select OKAY on the dialog box.**

RIBS

A Rib is a thin web feature created between features on a part (Figure 3–11). It can be revolved or straight. Ribs are similar to Protrusions that are extruded both sides from a sketching plane, with the exception that the section of a rib must be open. Additionally, the ends of a section of a rib must be aligned with existing part surfaces. Ribs sketched on a Through >> Axis datum plane and referenced to a surface of revolution will form a conical surface on the top of the rib (Figure 3–12).

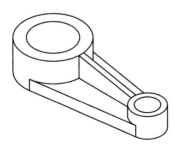

Figure 3-11   Rib on a part

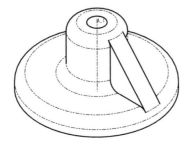

Figure 3-12   Rib on a revolved feature

CREATING A RIB

Figure 3–13 shows a part before and after a rib has been created. Perform the following steps to create this rib.

STEP 1: **Starting with the part shown in Figure 3–13, select FEATURE >> CREATE >> RIB.**

STEP 2: **Select a datum plane as the sketching plane for the rib (Figure 3–14).**

A Rib must be sketched on a datum plane. In Figure 3–14, datum plane FRONT intersects the middle of the two cylindrical features and is used in this example.

STEP 3: **Orient the sketching environment.**

STEP 4: **Within the sketcher environment, specify references to meet design intent (Figure 3–15).**

STEP 5: **Sketch the outline of the rib feature (Figure 3–16).**

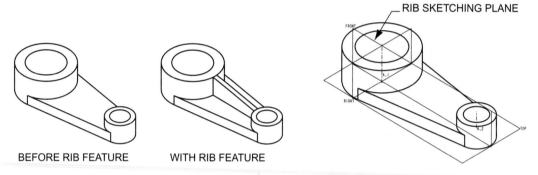

Figure 3-13   Creating a rib

Figure 3-14   Rib sketching plane

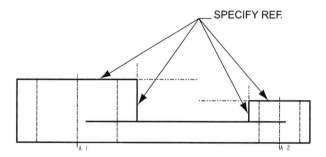

Figure 3-15   Specifying references for a rib

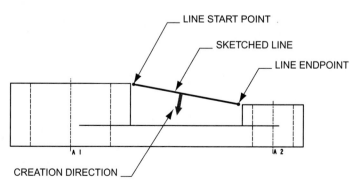

Figure 3-16   Sketching the rib

Sketch a line from the endpoints shown in Fig. 3–16. This will be the only entity required to define the section. Rib features must be sketched with an open section.

STEP 6: **Select the Continue option to exit the sketcher environment.**

STEP 7: **The direction of extrusion should point toward the part. If necessary, select FLIP on the direction menu (similarly to Figure 3–16), then select OKAY.**

STEP 8: **Enter an Extrusion thickness.**

This extrusion thickness functions similarly to the Both Sides option. The thickness provided will extrude both directions from the sketching plane.

COSMETIC FEATURES

Cosmetic features enhance the display of parts without complicated geometric features that require regenerations. Since cosmetic features are considered part features, they can be redefined and modified. Unlike geometric features, the line style defining a cosmetic feature (with the exception of cosmetic threads) can be modified using the Modify >> Line Style option. This option allows for the changing of line color, font, and style. Four cosmetic features are available: sketched, threads, grooves, user-defined, and ECAD Areas.

SKETCHED COSMETIC FEATURES

Sketched cosmetic features are useful for including names and logos on a part. These cosmetic features are sketched on a part surface using normal sketching environment techniques (Figure 3–17). They can be constructed with or without feature parameters. To create a nonparametric sketched cosmetic feature, delete the dimensions defining the feature before exiting the sketcher environment.

COSMETIC GROOVES

Cosmetic Grooves are sketched sections projected onto a part surface. The section is sketched with normal sketcher tools. The projected cosmetic feature has no defined depth. An example of a Cosmetic Groove is shown in Figure 3–17. In this example, the Cosmetic Groove was sketched on the same plane as the Cosmetic sketch and projected on to the receiving surfaces.

COSMETIC THREADS

Threads can be created within Pro/ENGINEER using the Helical option under the Protrusion and Cut commands. Despite this, often it is satisfactory to only symbolically display a thread. The Cosmetic Thread option allows for the creation of thread symbols that correspond to a simplified thread representation (Figure 3–18).

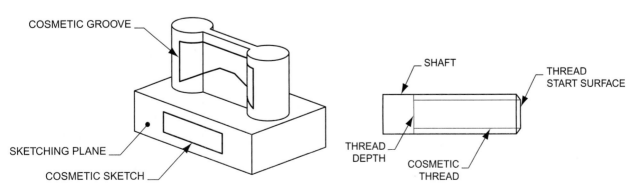

COSMETIC GROOVE

SKETCHING PLANE

COSMETIC SKETCH

SHAFT

THREAD START SURFACE

THREAD DEPTH

COSMETIC THREAD

Figure 3-17   Sketched cosmetic feature

Figure 3-18   Cosmetic threads

The parameters that define a cosmetic thread resemble parameters associated with true threads. As an example, internal and external threads can be defined as well as the major and minor diameters. The following are definable parameters for a cosmetic thread:

- **Major diameter** This parameter allows for the defining of the thread's major diameter. For external threads, the default value is 10 percent smaller than the thread surface diameter; for internal threads, the default value is 10 percent larger.

- **Threads Per inch** This parameter defines the number of threads-per-inch.

- **Thread form** This parameter allows for the selection of a thread form. An example would be the attribute UN (Unified National).

- **Thread class** The class of a thread is defined with this parameter. Examples include Fine, Extra Fine, and Coarse.

- **Placement** This parameter defines an external or internal thread.

- **Metric** This parameter sets the thread to metric (true) or not to metric (false).

CREATING A COSMETIC THREAD

Perform the following steps to create a cosmetic thread.

STEP 1: **Select FEATURE >> CREATE >> COSMETIC.**

Notice that the feature creation options include solid features, datum features, user-defined features, and cosmetic features.

STEP 2: **Select the THREAD option from the Cosmetic menu.**

Selecting the Thread option will display the Cosmetic Thread dialog box. This dialog box displays the parameters required for the creation of a cosmetic thread.

STEP 3: **Select the Thread Surface.**

On the work screen, select the cylindrical surface upon which to place the threads. Only one surface can be selected.

STEP 4: **Select the starting surface for the thread (Figure 3–18).**

The starting surface may be a part surface, quilt, or datum plane.

STEP 5: **Choose the direction of thread creation.**

A red arrow displays the default thread direction. Select OKAY to accept the direction or select Flip to change the direction.

STEP 6: **Select a thread depth specification, then provide appropriate depth information.**

Depth specifications include Blind, UpTo PntVtx, UpTo Curve, and UpToSurface. After selecting a specification, provide the necessary information that corresponds to the selected specification.

STEP 7: **Enter a thread diameter.**

The default diameter is displayed. For external threads, the diameter is 10 percent smaller than the thread surface diameter. For internal threads, the diameter is 10 percent larger than the thread surface diameter.

STEP 8: **Select one or more options from the Feature Parameters menu.**

The following options are available on the Feature Parameters menu:

- **Retrieve** This option allows for the retrieval of an existing thread parameter.

- **Save** This option allows a defined thread parameter to be saved for later use.

- **Mod Params** This option allows for the modification of a thread's parameters.
- **Show** This option shows parameters set for a thread.

STEP 9: Select DONE/RETURN on the Feature Parameters menu.

STEP 10: Select PREVIEW on the Cosmetic Thread dialog box, then select OKAY.

PATTERNED FEATURES

The Pattern command is used to create multiple instances of a feature. A feature is patterned by varying and duplicating one or more of its parametric dimensions. Instances of a pattern are copies of the feature. An instance is an exact duplicate of the parent feature. Two pattern configurations are available: linear and angular (Figure 3–19). One example of an angular pattern would be copying a hole around an axis, as with a bolt-circle pattern. Patterned features can be used effectively with dimensional relations to enhance the design intent of a model.

Two types of patterns are available: dimension patterns and reference patterns. With dimension patterns, varying one of the dimensions that defines the feature creates the new instances. For reference patterns, referencing a previously created pattern creates the new instances. The varying dimensions from the reference pattern govern the new pattern.

PATTERN OPTIONS

Three pattern options are available within Pro/ENGINEER. The decision of which option to use is based on the complexity of the part. Less complex patterns allow for more assumptions during the pattern construction. This allows Pro/ENGINEER to regenerate the pattern quicker. The following pattern options are available.

IDENTICAL PATTERNS

Identical patterns are the least complex and allow for the most assumptions. Identical pattern instances must be of the same size and must lie on the same placement surface. Identical pattern instances cannot intersect other features, instances, or the edge of the placement plane.

VARYING PATTERNS

Varying patterns are more complex than identical patterns. Varying pattern instances can vary in size and can lie on different placement surfaces. As with identical patterns, varying pattern instances cannot intersect other instances.

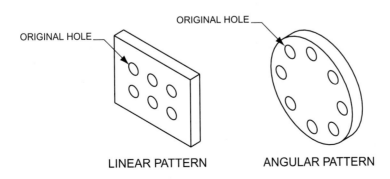

LINEAR PATTERN   ANGULAR PATTERN

Figure 3-19   Pattern configurations

GENERAL PATTERNS

General patterns are the most complex and require longer to regenerate. With general patterns, no assumptions are made during the construction process. Instances can intersect other instances and placement plane edges. Additionally, instances can vary in size and lie on different surfaces.

DIMENSIONS VARIATION

Dimensions can be varied using three options: Value, Relation, and Table. The following is a description of each option:

- **Value** With the value option, dimension values are incremented.
- **Relation** With the relation option, relations are used to drive dimension variations.
- **Table** With the table option, dimension variations are controlled by a table.

> **MODELING POINT**   When creating a feature that will be patterned, placement of dimensions defining the feature is critical. As an example, when creating a linear pattern, the dimensioning scheme defining the location of the feature will be used to create the pattern. For angular patterns, an angular dimension must exist.

CREATING A LINEAR PATTERN

Linear patterns are created by varying linear dimensions. A feature can be patterned unidirectionally or bidirectionally. Additionally, more than one dimension of a feature can be varied during the process. The following is a step-by-step approach for patterning the cylindrical feature shown in Figure 3–20. As shown, the 1.75-inch and 1.5-inch dimensions locating the feature will be patterned, as will the 2.0-inch and 1.25-inch dimensions defining the height and the diameter.

STEP   1: Select the FEATURE >> PATTERN command.

STEP   2: Select the feature to pattern.

On the work screen or on the Model Tree, select the cylindrical feature (Figure 3–20). With the Pattern command, only one feature can be patterned at a time.

STEP   3: Select VARYING >> DONE as the pattern option.

Other pattern options include Identical and General. Unlike the Identical option, the Varying option allows dimensional values on each pattern instance to vary. Also, unlike Identical, instances can lie on different placement surfaces.

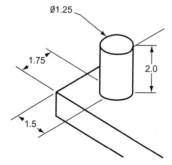

Figure 3-20   Feature to be patterned

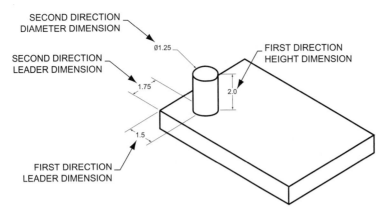

SECOND DIRECTION
DIAMETER DIMENSION

SECOND DIRECTION
LEADER DIMENSION

FIRST DIRECTION
HEIGHT DIMENSION

Ø1.25

1.75

2.0

1.5

FIRST DIRECTION
LEADER DIMENSION

Figure 3-21   Dimension selection

STEP 4: **Select VALUE on the Pattern Dimension Increment menu.**

Value is selected by default. Other options include Relation, Table, and Redraw Dims.

STEP 5: **Pick the 1.50-inch leader dimension for patterning in the first direction (Figure 3–21).**

On the work screen, select the first dimension to vary. This pattern will copy the feature in the first direction.

STEP 6: **Enter 2.75 as the dimension increment value.**

STEP 7: **(Optional Step) As shown in Figure 3–21, pick the height dimension for varying.**

The height value will be varied in the first direction.

STEP 8: **(Optional Step) Enter 0.25 as the dimension increment value.**

STEP 9: **Select DONE on the Exit menu.**

Selecting **Done** will end the definition of varying dimensions in the first direction.

STEP 10: **Enter 3 as the number of instances in the first direction.**

Three instances of the feature will be created in the first direction.

STEP 11: **Select VALUE from the Pattern Dimension Increment menu.**

The selection of a Dimension Variation option at this point will create a pattern in two directions. Selecting Done on the Exit menu would create a one-direction pattern.

STEP 12: **As shown in Figure 3–21, pick the 1.75 inch leader dimension for varying in the second direction.**

STEP 13: **Enter 2.50 as the dimension increment value.**

STEP 14: **(Optional Step) As shown in Figure 3–21, pick the cylinder diameter dimension for varying.**

You will also vary the diameter dimension in the second direction of the pattern.

STEP 15: **(Optional Step) Enter 0.500 as the dimension increment value.**

STEP 16: **Select DONE from the Exit menu.**

Selecting Done will end the selection of dimensions for varying.

STEP 17: **Enter 2 as the number of instances in the second direction.**

Two instances in the second direction in combination with three instances in the first direction will create a total of six instances. Due to the variations

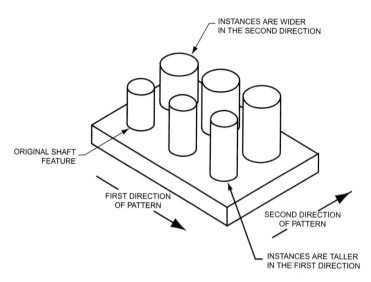

Figure 3-22   Pattern feature

selected for the height and the diameter dimensions of the cylinder, each instance of the pattern will be a different size.

STEP 18:   **Select DONE to create the pattern.**

After selecting Done, the pattern will appear as shown in Figure 3–22.

SUMMARY

While the Protrusion and Cuts commands are the primary tools to create features, other options are available that enhance the power of Pro/ENGINEER's modeling capabilities. As an example, if a part has multiple instances of a feature, it would be time consuming to individually model each feature. In this case, Pro/ENGINEER provides the Pattern command to create a rectangular or polar array of the feature. Feature creation tools such as Round, Chamfer, Shell, and Draft are available for a unique modeling situation. It is important to understand the capabilities of Pro/ENGINEER's various construction tools and to know when one is appropriate over another.

FEATURE CONSTRUCTION TUTORIAL 1

This tutorial exercise will provide instruction on how to model the part shown in Figure 3–23.

Within this tutorial, the following topics will be covered:

- Creating a new object.
- Creating an extruded protrusion.
- Creating a simple round.
- Creating a straight chamfer.
- Creating a standard coaxial hole.
- Creating an advanced round.

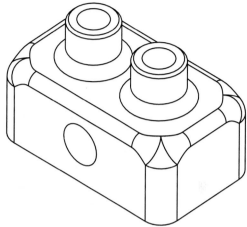

Figure 3-23   Finished model

STARTING A NEW OBJECT

This segment of the tutorial will establish Pro/ENGINEER's working environment.

STEP 1:   **Start Pro/ENGINEER.**

STEP 2:   **Set Pro/ENGINEER's Working Directory.**

From the File menu, select the Set Working Directory open; then establish an appropriate working directory for the part.

STEP 3:   **Start a NEW Pro/ENGINEER part and name it *construct1*.**

From the File menu, select New, then create a new part file named *construct1*. Use the default template file as supplied by Pro/ENGINEER. Your part model should start with a set of default datum planes, the default coordinate system, and *inch_lbm_second* as the units.

CREATING THE BASE GEOMETRIC FEATURE

This segment of the tutorial will create the extruded feature shown in Figure 3–24.

STEP 1:   **Select FEATURE >> CREATE >> PROTRUSION.**

This step will allow for the creation of a solid protrusion.

STEP 2:   **Select EXTRUDE >> SOLID >> DONE on the Solid Opts menu.**

STEP 3:   **Select ONE SIDE >> DONE on the Attributes menu.**

This feature will be extruded one direction from the sketching plane.

STEP 4:   **Select datum plane RIGHT as the sketching plane.**

STEP 5:   **Select OKAY on the Direction menu.**

Selecting OKAY will accept the default feature creation direction.

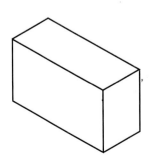

Figure 3-24   Extruded base feature

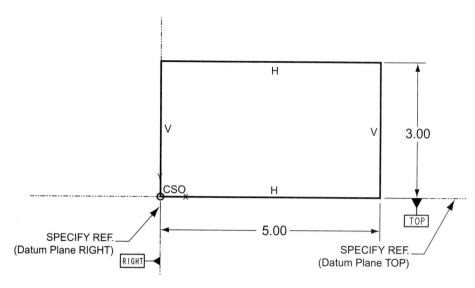

Figure 3-25   Regenerated section

STEP 6: **Select TOP on the Sketch View menu, then pick datum plane TOP on the work screen.**

This option will allow datum plane TOP to be oriented toward the top of the sketcher environment. The Default option would perform the same function.

STEP 7: **Close the References dialog box.**

When no existing geometric features exist for a part, Pro/ENGINEER will automatically specify datum planes RIGHT and FRONT as references. These are the only required references for this model.

STEP 8: Using the LINE icon, sketch the section shown in Fig. 3–25.

As shown in Figure 3–25, sketch the Line entities aligned with datum planes RIGHT and TOP. When sketching, do not worry about the size of the entities being sketched. Entity size parameters will be modified in another step. Allow Intent Manager to apply the vertical and horizontal constraints as shown.

> **MODELING POINT** When sketching, your left mouse button is used to pick entity points on the work screen and your middle mouse button is used to cancel commands.

STEP 9: Use the MODIFY icon to match the dimension values shown in Figure 3–25.

Select Modify then select the dimension value to change. Follow this by entering a new value for the dimension.

STEP 10: Select the Continue icon when the section is complete.

STEP 11: **Select BLIND >> DONE as the depth option.**

STEP 12: **Enter 2.00 as the extrude depth.**

STEP 13: **Preview the feature.**

After selecting the Preview option, shade and dynamically rotate the object. It is not too late to change a parameter defining the feature. Selecting the parameter on the Feature Definition dialog box followed by the Define option will allow the selected parameter to be modified. As an example, the depth of the extrusion can be changed by redefining the depth parameter.

STEP 14: **If the Protrusion is correct, select OKAY on the Feature Definition dialog box.**

ADDING EXTRUDED FEATURES

An Extruded Protrusion can be sketched on an existing part surface. This section of the tutorial will create the Protrusion shown in Figure 3–26.

STEP 1: **Select FEATURE >> CREATE >> PROTRUSION.**

STEP 2: **Select EXTRUDE >> SOLID >> DONE on the Solid Options menu.**

STEP 3: **Select ONE SIDE >> DONE.**

STEP 4: **Select the top surface of the part as the sketching plane (Figure 3–26).**

As shown in Figure 3–26, select the top surface of the existing part to use as a sketching plane. Observe the dimensions of the part; care should be taken not to select one of the side surfaces.

STEP 5: **Select OKAY to accept the default feature creation direction.**

STEP 6: **Select DEFAULT to accept the default sketch orientation.**

STEP 7: **Close the References dialog box.**

STEP 8: On the toolbar, turn off the display of datum planes.

STEP 9: On the toolbar, select No Hidden as the model display.

Sketching a new feature on an existing part surface is usually easier with No Hidden or Hidden set as the model display style.

STEP 10: Using the CIRCLE icon, sketch the two circles shown in Figure 3–27.

The Circle option defines a circle by selecting the circle's center point, then dragging the size of the circle. Allow Intent Manager to create a Horizontal Alignment constraint between the centers of the two circles.

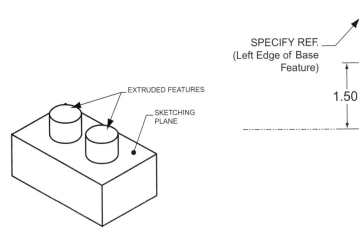

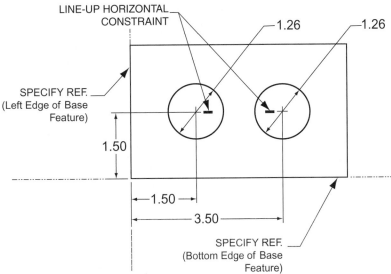

Figure 3-26   Extruded features

Figure 3-27   Sketched regenerated section

STEP 11: Use the DIMENSION icon to match the dimensioning scheme shown in Figure 3–27.

To dimension the location of a circle, with the left mouse button, select the center of the circle, then the locating edge. Place the dimension with the right mouse button. Diameter dimensions are created by double picking the circle with the left mouse button, followed by placing the dimension with the right mouse button.

> **MODELING POINT**   An alternative to the two separate diameter dimension shown in Figure 3–27 would be to create an equal radii constraint between the two circles. With an equal radii constraint, one dimension will control the size of both circles. This is a common way to capture design intent. An equal radii constraint can be created through the Constraint icon and the Equal option.

STEP 12: Using the Select icon, double pick dimensions and modify their values to match Figure 3–27.

STEP 13: Select the Continue icon on the Sketch menu when the section is complete.

STEP 14: Enter a BLIND depth of 1.00.

Select Blind as the depth option, then enter a value of 1.00.

STEP 15: Preview the feature.

STEP 16: Select OKAY from the dialog box to create the feature.

CREATING ROUNDS

This segment of the tutorial will create the simple rounds shown in Figure 3–28.

STEP 1: Select FEATURE >> CREATE >> ROUND.

The Round command is available to create rounds and fillets. A round is considered a part feature.

STEP 2: Select SIMPLE >> DONE on the Round Type menu.

STEP 3: Select CONSTANT >> EDGE CHAIN >> DONE on the Round Set Attributes menu.

The Constant option will create a round with a constant radius while the Edge chain option will allow for the selection of edges to round.

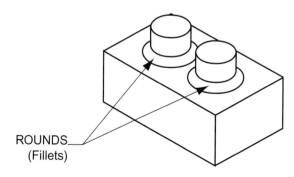

ROUNDS
(Fillets)

Figure 3-28   Part with simple rounds

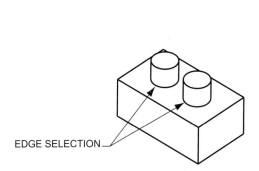

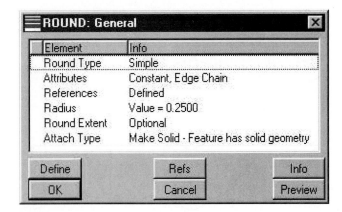

Figure 3-29   Round edge selection

Figure 3-30   Round Feature Definition dialog box

STEP 4:   **Select ONE BY ONE >> DONE on the Chain menu.**

The One-By-One option requires the individual selection of each edge.

STEP 5:   **As shown in Figure 3–29, select the two edges at the base of the circular protrusion.**

STEP 6:   **Select DONE on the Chain menu.**

Select Done on the chain menu when you are through selecting edges to round.

STEP 7:   **Enter .25 as the radius of the rounds.**

STEP 8:   **Preview the rounds, then select OKAY on the Round Feature Definition dialog box (Figure 3–30).**

CREATING A CHAMFER

Pro/ENGINEER provides the option to create either a corner chamfer or an edge chamfer. This segment of the tutorial will create the edge chamfer shown in Figure 3–31.

STEP 1:   **Select the FEATURE >> CREATE >> CHAMFER command.**

STEP 2:   **Select EDGE as the type of chamfer to create.**

The Edge option will create an angled surface between two planes. The Corner option will create an angled surface between three planes.

STEP 3:   **On the Scheme menu, select 45 × d (Figure 3–32).**

The 45 × d option will create a 45-degree angle with a user-defined distance.

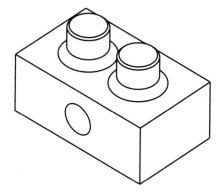

Figure 3-31   Edge chamfer

Figure 3-32   Chamfer schemes

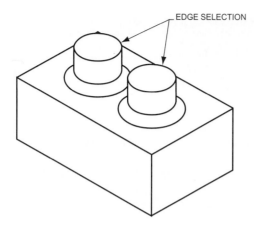

EDGE SELECTION

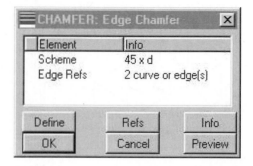

Figure 3-33   Edge selection

Figure 3-34   Chamfer Feature Definition dialog box

STEP 4: Enter .0625 as the Chamfer Dimension.

STEP 5: As shown in Fig. 3–33, select the edges on the top of the two circular protrusions.

STEP 6: Select DONE SEL on the Get Select menu.

When you are through selecting edges to round, use the Done Sel option to exit the Get Select menu.

STEP 7: Select DONE REFS on the Feature Refs menu.

The Feature Refs menu allows for the addition or removal of edges from the edge selection set.

STEP 8: From the Feature Definition dialog box, select Preview; then select OKAY (Figure 3–34).

CREATING A STANDARD COAXIAL HOLE

This segment of the tutorial will create the two coaxial threaded holes shown in Figure 3–35. The specification will be a standard 0.75 inch unified national course internal thread.

STEP   1: Select FEATURE >> CREATE >> HOLE.

STEP   2: On the Hole dialog box, select STANDARD as the Hole Type.

Standardized thread specification tables are used to define standard holes. As an example, a 3/4-inch nominal size hole with unified national course threads has 10 treads per inch. Pro/ENGINEER has three preexisting hole

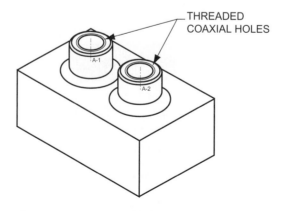

THREADED
COAXIAL HOLES

Figure 3-35   Coaxial holes

Figure 3–36   Hole type

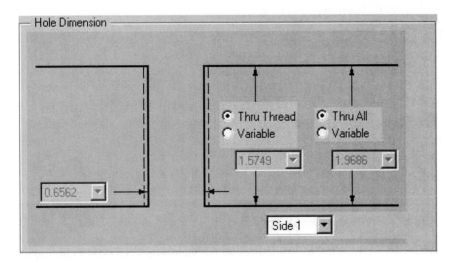

Figure 3–37   Hole depth options

tables: UNC, UNF, and ISO. Each table provides parameters needed to detail a threaded hole.

STEP   3:   On the Hole dialog box, set the options shown in Figure 3–36.

Set the following options: Tapped Hole, 3/4-10 Screw Size, and Add Thread Surface. Ensure that the Add Counterbore and the Add Countersink options are unchecked.

STEP   4:   Set THRU THREAD and THRU ALL as shown in Figure 3–37.

The Thru All option will construct the hole through the entire part. Similarly, the Thru Thread option will construct the thread through the entire length of the hole. Notice in the figure the dimmed out 0.6562 value. This value represents the tap drill size. The hole within your part model will be constructed with this diameter value. This value is obtained from the UNC hole chart.

STEP   5:   As shown in Figure 3–38, select the first axis as the Primary Reference.

STEP   6:   As shown in Figure 3–38, select the holes placement plane.

STEP   7:   On the dialog box, select the Preview icon.

STEP   8:   Select the Build Feature checkmark.

STEP   9:   On the menu bar, select UTILITIES >> ENVIRONMENT and uncheck the 3D NOTES option.

Deselecting the 3D Notes environmental option will turn off the display of the standard hole's thread representation note. This option can also be set with the configuration file option *model_note_display*.

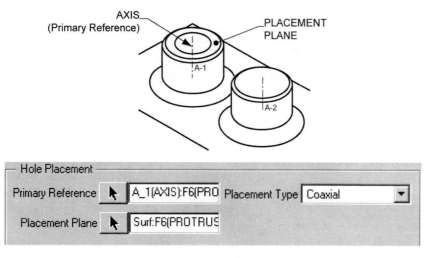

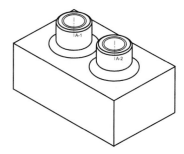

Figure 3-38   Hole placement references

Figure 3-39   Two coaxial holes

STEP 10:   Repeat steps 1 through 8 to create the second standard coaxial hole (Figure 3–39).

STEP 11:   Save your part.

CREATING A LINEAR HOLE

The Linear Hole placement option will locate a hole from two reference edges. This tutorial will create the linear hole shown in Figure 3–40.

STEP   1:   Select FEATURE >> CREATE >> HOLE.

STEP   2:   Select STRAIGHT as the Hole Type.

STEP   3:   Enter 1.000 as the hole's Diameter (see Figure 3–41).

STEP   4:   Select THRU ALL as the Depth One parameter. (Leave Depth Two set at None.)

STEP   5:   Select the front surface of the part as the Primary Reference (see Figure 3–42).

STEP   6:   Pick the first Linear Reference edge for locating the hole (Figure 3–42).

The reference edges shown in Figure 3–42 are used to locate the hole. Locating the hole from these edges matches the intent of the design.

STEP   7:   In the Hole dialog box, enter a distance of 2.50 for the first Linear Reference (Figure 3–41).

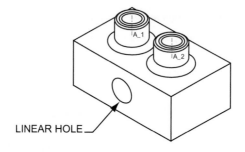

LINEAR HOLE

Figure 3-40   Linear hole placement

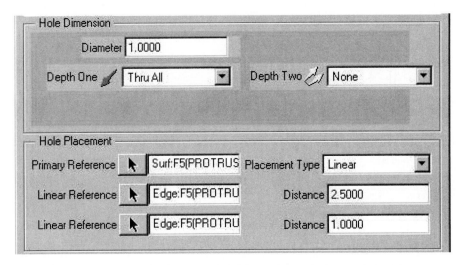

Figure 3-41   Straight hole parameters

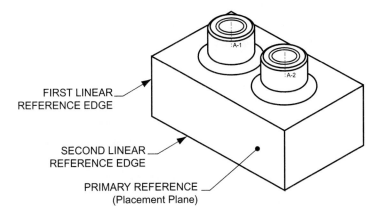

Figure 3-42   Hole references

DESIGN INTENT   When utilizing the Linear placement option and the placement plane shown in Figure 3-42, any edge of the placement plane could be used to locate the hole. Since the design intent requires the hole to be located from the edges shown in the figure, these references should be the edges selected during part modeling. Pro/ENGINEER is a Feature-Based/Parametric modeling system. These types of systems allow for the incorporation of design intent into a model. Placing a hole is one example of where design intent can be incorporated. In this example, placing the hole from these edges incorporates design intent.

STEP 8: Pick the second Linear Reference edge for locating the hole, then enter a distance of 1.00.

STEP 9: Select the Build Feature checkmark to create the hole.

STEP 10: Save your part.

CREATING AN ADVANCED ROUND

The Advanced Round created in this segment of the tutorial will consist of two Round Sets. The first round set will have a radius value of 0.500, while the second round set will have a radius value of 0.235. The result of this round feature is shown in Figure 3–43.

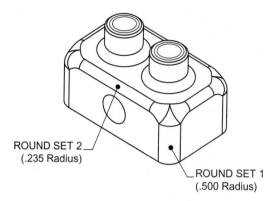

ROUND SET 2
(.235 Radius)

ROUND SET 1
(.500 Radius)

Figure 3-43   Advanced Round feature

STEP   1:   **Select FEATURE >> CREATE >> ROUND.**

STEP   2:   **Select ADVANCED >> DONE on the Round Type menu.**

Advanced rounds allow for varied cross-sections and multiple round sets.

STEP   3:   **Select ADD on the Round Sets menu.**

The Add option will add a Round set to the round definition.

STEP   4:   **Select CONSTANT >> EDGE CHAIN >> DONE on the Round Set Attribute menu.**

STEP   5:   **Select ONE BY ONE on the Chain menu.**

The One-by-One option will require you to select each individual edge.

STEP   6:   **As shown in Figure 3–44, pick the four vertical corners of the part.**

INSTRUCTIONAL NOTE   The fourth corner is hidden in Figure 3–44. Use the Query Sel option to pick this edge or dynamically rotate the part.

STEP   7:   **After the four corners are selected, choose DONE on the Chain menu.**

STEP   8:   **Enter .500 as this round set's radius value.**

After entering the radius value, observe the Round Set 1 dialog box (Figure 3–45).

STEP   9:   **Select OKAY on the ROUND SET 1 Feature Definition dialog box.**

Selecting OKAY will accept the parameters defined for Round Set 1. Before selecting OKAY, each parameter can be redefined with the Define option.

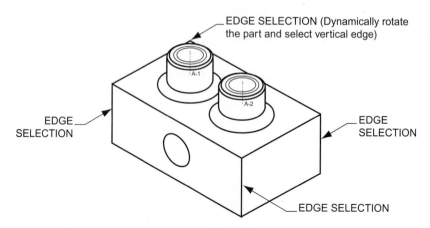

EDGE SELECTION (Dynamically rotate
the part and select vertical edge)

EDGE
SELECTION

EDGE
SELECTION

EDGE SELECTION

Figure 3-44   Edge selection

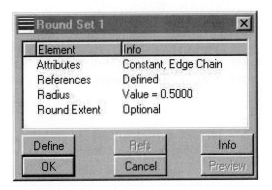

Figure 3-45   Round set one

Step 10:   Select ADD on the Round Sets menu.

Selecting the Add option will create another round set.

Step 11:   Select CONSTANT >> EDGE CHAIN >> DONE.

Step 12:   Select ONE BY ONE on the Chain menu.

Step 13:   As shown in Figure 3–46, pick the four edges that define the top of the part.

Step 14:   Select DONE on the Chain menu.

Step 15:   Enter .235 as this round set's radius value.

Step 16:   Select OKAY on the ROUND SET 2 Feature Definition dialog box (Figure 3–47).

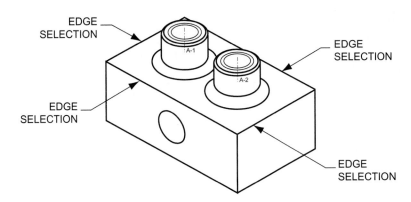

Figure 3-46   Edge selection

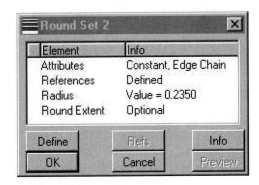

Figure 3-47   Round set two

STEP 17: Select DONE SETS on the Round Sets menu.

Select Done Sets when you are through creating round sets.

STEP 18: Select PREVIEW on the Round Feature definition dialog box.

STEP 19: Finish the round by selecting OKAY on the dialog box.

INSERTING A SHELL

This segment of the tutorial will create the shelled feature shown in Figure 3–48. The Shell command removes selected surfaces from a part and provides the remaining surfaces with a user-defined wall thickness.

STEP 1: Select FEATURE >> CREATE >> SHELL.

STEP 2: Select the Surface Shown in Figure 3–49.

The selected surface will be removed from the part.

STEP 3: Select DONE SEL on the Get Select menu.

STEP 4: Select DONE REFS on the Feature Refs menu.

The Feature Refs menu allows for the addition or removal of part surfaces.

STEP 5: Enter .125 as the Wall Thickness.

STEP 6: Preview the Shelled feature, then select OKAY on the Shell Feature Definition dialog box.

STEP 7: Save your part.

Figure 3-48 Shelled feature

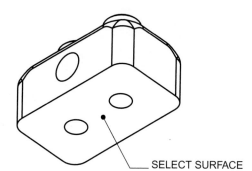

SELECT SURFACE

Figure 3-49 Select shell surface

FEATURE CONSTRUCTION TUTORIAL 2

This tutorial exercise will provide instruction on how to model the part shown in Figure 3–50.

Within this tutorial, the following topics will be covered:

- Creating a new object.
- Creating an offset datum plane.
- Creating an extruded protrusion.
- Creating a straight coaxial hole.
- Creating a linear-straight hole.
- Creating a linear pattern.
- Creating a straight chamfer.
- Creating a cut feature.
- Creating a rib.
- Creating a simple round.

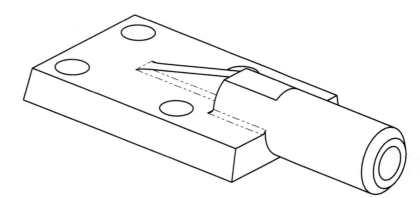

Figure 3-50   Finished model

STARTING A NEW PART

This segment of the tutorial will establish Pro/ENGINEER's working environment.

STEP 1:   **Start Pro/ENGINEER.**

STEP 2:   **Establish an appropriate working directory.**

STEP 3:   **Create a new part file with the name *CONSTRUCTION2* (Use Pro/ENGINEER's Default Template).**

CREATING THE BASE GEOMETRIC FEATURE

Model the Base Geometric Feature. The Base Feature is constructed from an 8 by 5-inch rectangular section extruded one direction a blind distance of 1 inch. Sketch the section on datum plane FRONT. Align the middle of the part with datum plane TOP and the left edge of the part with datum plane RIGHT. Figure 3–51 shows the finished feature and the sketched section.

CREATING AN OFFSET DATUM PLANE

This section of the tutorial will create an offset datum plane. As shown in Figure 3–52, the datum plane will be offset from datum plane RIGHT a distance of 6 inches. A datum plane may be offset from an existing part planar surface or an existing datum plane.

STEP 1:   Select the **CREATE DATUM PLANE** icon on the Datum toolbar.

STEP 2:   **Select OFFSET on the Datum Plane menu.**

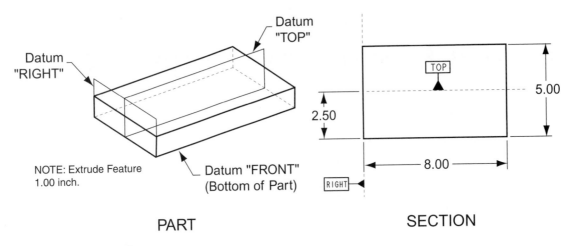

PART   SECTION

Figure 3-51   Base feature

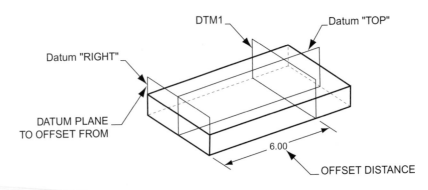

Figure 3-52   Datum offset

STEP   3: **As shown in Figure 3–52, select datum plane RIGHT as the surface to offset from.**

On the work screen or on the model tree then select datum plane RIGHT.

STEP   4: **On the Offset Menu, select the ENTER VALUE option.**

Pro/ENGINEER provides two options for setting the offset of a datum plane. The default option, Thru Point, will create the offset datum plane through an existing datum point. The Enter Value option requires you to enter an offset distance.

After selecting the Enter Value option, on the work screen, notice how Pro/ENGINEER graphically displays the direction of offset (Figure 3–53). To offset in the opposite direction, enter a negative value.

STEP   5: **In Pro/ENGINEER's textbox, enter a value that will offset the new datum plane 6 inches in the direction shown in Figure 3–53.**

A positive or negative value can be entered. If your offset direction is in the opposite direction, enter a negative value.

STEP   6: **Select DONE on the Datum Plane menu.**

Notice on the work screen and on the model tree the addition of datum plane DTM1. Pro/ENGINEER names datum planes in sequential order starting with DTM1. The next two steps in the tutorial will rename datum plane DTM1.

STEP   7: **Select SET UP >> NAME from Pro/ENGINEER's Menu Manager.**

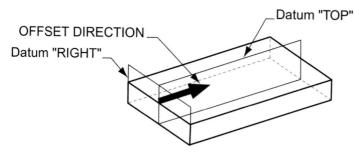

Datum "TOP"

OFFSET DIRECTION

Datum "RIGHT"

Figure 3-53   Datum offset direction

Figure 3-54   Datum plane selection

Step   8:   On the Model Tree, select the datum plane DTM1 feature
(Figure 3–54).

Step   9:   In the textbox enter *OFFSET_DATUM* as the new name for datum
plane DTM1.

Notice the underscore between *Offset* and *Datum*. On the Model Tree, new
feature names cannot have spaces.

MODELING POINT   The default names that Pro/ENGINEER gives features are not descriptive. It is
helpful to rename them to allow for ease of identification and selection.

Step 10:   Save your part.

Creating an Extruded Protrusion

This section of the tutorial will create the cylindrical protrusion shown in Figure 3–55. This
protrusion will consist of a circle entity sketched on the offset datum plane (*Offset_Datum*).
The section will be extruded a depth of 6 inches.

Step 1:   Setup a ONE SIDE Extruded Protrusion with datum plane
OFFSET_DATUM as the sketching plane and the extrude direction set as
shown in Figure 3–56.

Setup this feature as a One-Sided extruded Protrusion. Select datum plane
Offset_Datum as the sketching plane (Figure 3–56). Within Pro/ENGINEER,
there are two sides to a datum plane: the positive side and the negative side.
Using Pro/ENGINEER's default color scheme (see the Color option under the
Utilities menu), the positive side of a datum plane is yellow and the negative
side is red. When selecting a datum as the sketching plane for an extruded

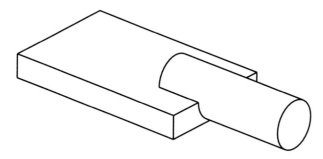

Figure 3-55   Extruded protrusion

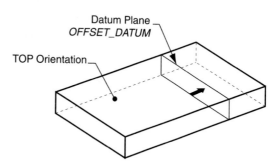

Datum Plane
OFFSET_DATUM

TOP Orientation

Figure 3-56   Extrude direction

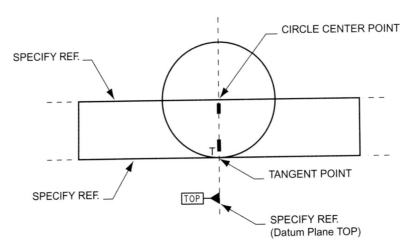

Figure 3-57   Feature's section

protrusion, by default the feature will be extruded in the positive direction. To extrude in the negative direction, used the Direction menu's Flip option.

STEP 2: **Orient the sketching environment as shown in Figure 3–57.**

Use the Top option and select the top of the part.

STEP 3: ⬜ **On the Model Display toolbar, select NO HIDDEN as the model's display.**

STEP 4: ▱ **On the Datum Display toolbar, turn off the display of datum planes.**

STEP 5: **Using the References dialog box, specify the three references shown in Figure 3–57.**

STEP 6: ⭕ **Sketch a Circle entity as shown in Figure 3–57.**

The center of the circle is aligned with the top of the part and with datum plane TOP. The bottom edge of the circle is aligned with the bottom of the part (creating a Tangent constraint).

> **DESIGN INTENT**   The design intent for this extruded feature requires the diameter of the cylindrical feature to be twice the thickness of the base feature. Also, the design intent requires the center of the feature to be aligned with the top of the part. By sketching the feature in the manner shown in Figure 3–57, the intent of the design will be captured. For this part, if the thickness of the base feature is changed (it is currently set at 1 inch), the size of the cylindrical feature will change accordingly. This meets the intent for this design.

STEP 7: **Select the Continue icon to exit the sketcher environment.**

STEP 8: **Construct a BLIND depth of 6 inches.**

STEP 9: **Preview the feature, then select OKAY on the Extrude Feature Definition dialog box.**

CREATING A COAXIAL HOLE

This segment of the tutorial will create the Straight-Coaxial Hole shown in Figure 3–58.

STEP 1: **Select FEATURE >> CREATE >> HOLE.**

STEP 2: **On the Hole dialog box, select STRAIGHT as the Hole Type.**

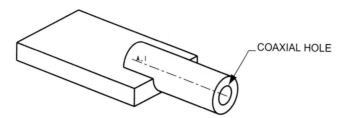

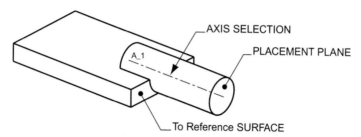

Figure 3-58   Coaxial hole

COAXIAL HOLE

AXIS SELECTION

PLACEMENT PLANE

To Reference SURFACE

Figure 3-59   Coaxial hole references

STEP 3: Enter 1.00 as the diameter for the hole.

STEP 4: Select TO REFERENCE as the Depth One parameter, then select the To Reference Surface shown in Figure 3–59.

STEP 5: Pick the axis of the previously created protrusion feature (Figure 3–59) as the Primary Reference for the hole.

STEP 6: Pick the end of the protrusion feature as the Placement Plane (Figure 3–59).

STEP 7: Preview the hole.

STEP 8: Select the Build Feature checkmark.

CREATING A LINEAR HOLE

This section of the tutorial will create the Straight-Linear Hole shown in Figure 3–60.

STEP 1: Select FEATURE >> CREATE >> HOLE.

STEP 2: Select STRAIGHT as the Hole Type.

STEP 3: Enter 1.00 as the Hole's Diameter.

STEP 4: Select THRU ALL as the Depth One parameter.

STEP 5: Pick the Placement Plane, shown in Figure 3–61, as the PRIMARY REFERENCE.

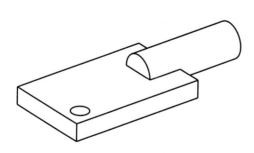

Figure 3-60   Straight-linear hole

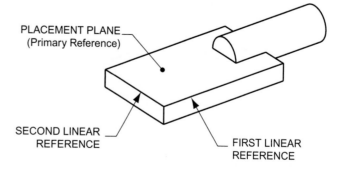

PLACEMENT PLANE
(Primary Reference)

SECOND LINEAR REFERENCE

FIRST LINEAR REFERENCE

Figure 3-61   Linear hole references

Step 6: Select the First LINEAR REFERENCE (see Figure 3–61), then enter a value of 1.00 for the reference's distance.

Step 7: Select the Second LINEAR REFERENCE, then enter a value of 1.00 for the reference's distance.

Step 8: Select the Build Feature checkmark.

CREATING A LINEAR PATTERN

This segment of the tutorial will create a linear pattern of the hole created in the previous section of this tutorial. The Pattern command allows for the copying of a feature in one or two directions. To pattern a feature, a leader dimension is required. In this example, the two linear dimensions used to locate the parent hole will serve as the leader dimensions used. Figure 3–62 shows the finished pattern.

Step 1: Select the FEATURE >> PATTERN command.

Step 2: Select the Straight-Linear hole created in the previous section of this tutorial.

On the work screen or on the model tree, select the previously created Hole feature.

Step 3: Select IDENTICAL as the Pattern option.

The Identical option requires the most assumptions. The following assumptions must be met:

- The pattern must be on one placement surface.
- The pattern cannot intersect an edge.
- The pattern's instances cannot intersect.

Step 4: As shown in Figure 3–63, pick the First Leader dimension for patterning in the First Direction.

Features can be patterned in two directions. In this step, you are selecting a leader dimension that will be used to determine the first direction of pattern. Additionally, this leader dimension will be used to determine the distance between each instance of the pattern.

Step 5: Enter 4.00 as the dimension increment value.

Each instance of the pattern in the first direction will be 4 inches apart.

Step 6: Select DONE on the Exit menu, then enter 2 as the number of instances.

In addition to the leader dimension, multiple dimensions can be varied in a direction of pattern. As an example, the hole's diameter can be varied with each instance of the pattern. In this tutorial, no additional dimensions will be varied.

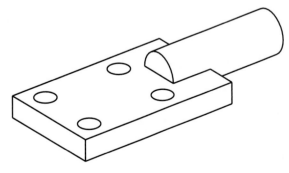

Figure 3-62   Patterned hole

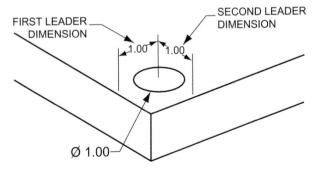

Figure 3-63   Dimension selection

STEP 7: As shown in Figure 3–63, select the Second Leader dimension for
patterning in the Second Direction.

This tutorial will create a pattern in two directions. Select the second hole
placement dimension as shown in Figure 3–63.

STEP 8: Enter 3.00 as the dimension increment value.

Each instance of the pattern in the second direction will be 3 inches apart.

STEP 9: Select DONE on the Exit menu, then enter 2 as the number of
instances.

The pattern will be created after entering the number of instances.

STEP 10: Save the part.

CREATING A CHAMFER

This segment of the tutorial will create the Straight Chamfer shown in Figure 3–64.

STEP 1: Select FEATURE >> CREATE >> CHAMFER.

STEP 2: Select EDGE as the chamfer type.

STEP 3: Select 45 × d as the chamfer's dimensioning scheme.

The 45 × d option will create a chamfer that consists of a 45-degree angle
with a user-specified distance.

STEP 4: Enter .25 as the chamfer's dimension value.

STEP 5: Pick the feature edge as shown in Figure 3–65, then select DONE SEL on
the Get Select menu.

STEP 6: Select DONE REFS on the Feature Reference menu.

The Feature Reference menu allows for the selection or removal of edges to
chamfer.

STEP 7: Preview the chamfer, then select OKAY on the Feature Definition dialog
box.

STEP 8: Save the part.

Figure 3-64 Chamfer feature   **Figure 3-65** Chamfer edge selection

CREATING A CUT

This section of the tutorial will create the Cut feature shown in Figure 3–66.

STEP 1: Select FEATURE >> CREATE >> CUT.

STEP 2: Setup a One-Sided Extruded Cut with the sketching plane and
orientation shown in Figure 3–67.

STEP 3: In the sketching environment, on the Model Display toolbar select
NO HIDDEN as the model's display style.

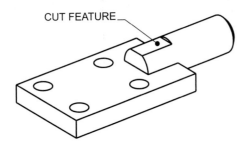

Figure 3-66   Cut feature

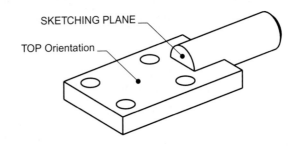

Figure 3-67   Cut references

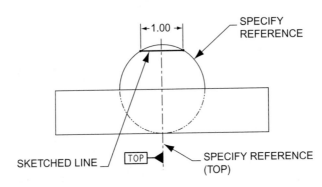

Figure 3-68   Cut section

STEP 4:   On the Datum Display toolbar, turn off the display of datum planes.

STEP 5:   Using the References dialog box, pick the edge of the circular protrusion as a reference for this section (see Figure 3–68).

STEP 6:   Sketch the Line entity shown in Figure 3–68, then use the DIMENSION and MODIFY options to create the dimensioning scheme shown.

STEP 7:   Select the Continue icon to exit the Sketcher menu, then accept the default material removal side.

STEP 8:   Create a BLIND depth of 2.00.

STEP 9:   Preview the feature, then select OKAY on the Feature Definition dialog box.

CREATING A RIB

This segment of the tutorial will create the Rib feature shown in Figure 3–69. A Rib feature is similar to an Extruded Protrusion. With Ribs, the section has to be opened and the extrude direction is both-sides by default. In this tutorial, the Rib will be sketched on Datum Plane TOP.

STEP 1:   Select FEATURE >> CREATE >> RIB.

STEP 2:   Select Datum Plane TOP as the sketching plane (Figure 3–70).

Datum plane TOP can be picked on the work screen or on the Model Tree. For parts with multiple features, it is often easier to use the Model Tree to select a specific feature.

STEP 3:   Orient the top of the part toward the top of the work screen (Figure 3–70).

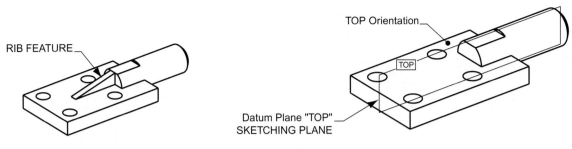

Figure 3-69   Rib feature   **Figure 3-70**   Orientation selection

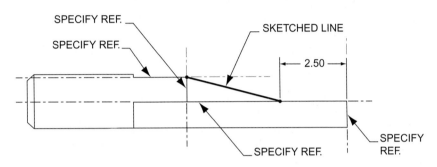

Figure 3-71   Sketched section

STEP 4: Using the References dialog box, specify the four references shown in Figure 3–71.

STEP 5: Sketch the single LINE shown in Figure 3–71.

STEP 6: Use the DIMENSION and MODIFY options to create the dimensioning scheme shown in Figure 3–71.

With the correct references specified, only one dimension is needed to fully define the section. In this example, the design intent requires the end of the rib to be measured from the end of the part.

STEP 7: Select the Continue icon to exit the sketcher environment.

STEP 8: If necessary, select FLIP >> OKAY to change the direction of material creation.

Due to the open section, you have to specify the side of the sketch on which to create material. Your arrow should point toward the direction where material will be added.

STEP 9: Enter .500 as the Rib's thickness.

Pro/ENGINEER will create the feature after entering the Rib's thickness. Notice how the Rib command does not utilize a Feature Definition dialog box.

CREATING A DRAFT

This section of the tutorial will draft the three surfaces shown in Figure 3–72.

STEP   1: Select FEATURE >> CREATE >> TWEAK.

STEP   2: Select DRAFT on the Tweak menu.

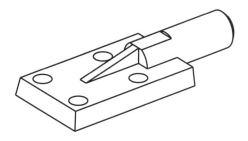

Figure 3-72 Drafted surfaces

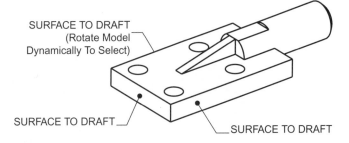

SURFACE TO DRAFT
(Rotate Model
Dynamically To Select)

SURFACE TO DRAFT

SURFACE TO DRAFT

Figure 3-73 Surface selection

STEP 3: Select NEUTRAL PLN >> DONE as the Draft option.

Draft options available include Neutral Plane for pivoting around a plane or surface and Neutral Curve for pivoting around a datum curve or edge.

STEP 4: Select TWEAK >> NO SPLIT >> CONSTANT >> DONE on the Draft Attributes menu.

No Split will create a draft on the complete side of a selected surface. The Split at Plane option will create a split surface at the neutral plane, while the Split at Sketch option will create a draft out of a user-sketched surface.

STEP 5: Select surfaces to draft (Figure 3–73).

Select the three surfaces shown in Figure 3–73. You will need to rotate the model dynamically or use the Query Sel option to select the third surface.

STEP 6: Select DONE SEL on the Get Select menu.

Selecting Done Sel will return you to the Surface Select menu. The Surface Select menu provides you with options for adding surfaces, excluding surfaces, and showing existing selected surfaces.

STEP 7: Select SHOW >> MESH on the Surface Select menu.

The **Show >> Mesh** option will show selected surfaces as a mesh (Figure 3–74).

STEP 8: Select DONE on the Surface Select menu.

STEP 9: Select the bottom of the part as the Neutral Plane (Figure 3–75).

The Neutral Plane is the part surface that will remain intact during the drafting procedure. Since the drafted surfaces will pivot from the neutral plane, this is a critical selection.

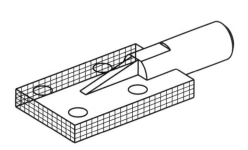

Figure 3-74 Show draft surfaces

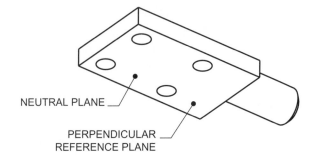

NEUTRAL PLANE

PERPENDICULAR
REFERENCE PLANE

Figure 3-75 Neutral plane and reference plane selection

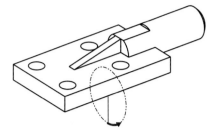

Figure 3-76   Draft rotation reference

STEP 10: **Select the Bottom of the part as the perpendicular reference plane (Figure 3–75).**

The reference plane has to lie perpendicular to the draft surfaces. Often, this reference plane is the same as the neutral plane.

STEP 11: **Enter a draft angle that will create a 10-degree draft in the direction shown in Figure 3–76.**

As shown in Figure 3–76, Pro/ENGINEER graphically displays the draft's rotation direction. The rotation direction should be as shown in the figure. If necessary, a negative value can be entered to reverse this direction.

STEP 12: **Preview the Draft, then select OKAY on the Feature Definition dialog box.**

CREATING A ROUND

This segment of the tutorial will create the rounds shown in Figure 3–77. Despite their apparent simplicity, rounds can be one of the most troublesome commands in Pro/ENGINEER. Since it is advisable not to use a round as a reference for the creation of a feature, it is recommend that rounds be created as the last features of a part.

STEP 1: **Select FEATURE >> CREATE >> ROUND.**

STEP 2: **Select SIMPLE >> DONE as the Round type to create.**

STEP 3: **Select CONSTANT >> EDGE CHAIN >> DONE as the Round Set Attributes.**

STEP 4: **Select ONE BY ONE as the Chain selection option.**

STEP 5: **Select the Edges shown in Figure 3–78 and the corresponding edges on the opposite side of the Rib and Cylinder.**

On the work screen, select the edges shown in Figure 3–78. Use the Query Sel option to select the corresponding edges on the opposite side of the rib and cylinder features.

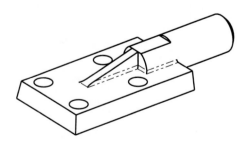

Figure 3-77   Rounded features

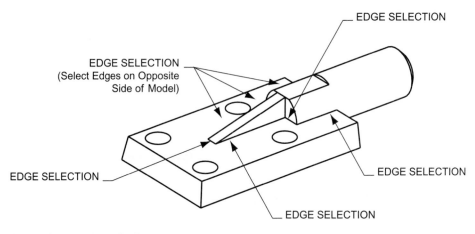

Figure 3-78   Round selection

STEP 6: Select DONE on the Chain menu then enter a round radius value of .125.

STEP 7: Preview the rounds on the Feature Definition dialog box, then select OKAY.

STEP 8: Save your part.

STEP 9: Purge the old part file by using the FILE >> DELETE >> DELETE OLD option.

Every time a file is saved in Pro/ENGINEER, a new version of the file is created. The Delete Old option will delete old versions.

PROBLEMS

1.   Using Pro/ENGINEER's Part mode, model the part shown in Figure 3–79. Construct the part using the following order of operation:

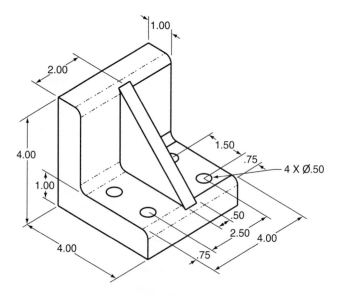

ALL ROUNDS AND FILLETS R.25

Figure 3-79   Problem one

a.   Use Pro/ENGINEER's default template file.

b.   Sketch the base feature on Datum Plane FRONT.

c.   Model the Base Protrusion as a Both-Sides extrusion.

d.   Construct the Rounded features.

e.   Model the first hole as a Linear-Straight Hole feature.

f.   Pattern the Hole feature.

g.   Construct the Rib feature by sketching on Datum Plane FRONT.

2.   Using Part mode, model the part shown in Figure 3–80.

3.   Use Pro/ENGINEER to model the part shown in Figure 3–81.

4.   Model the part shown in Figure 3–82.

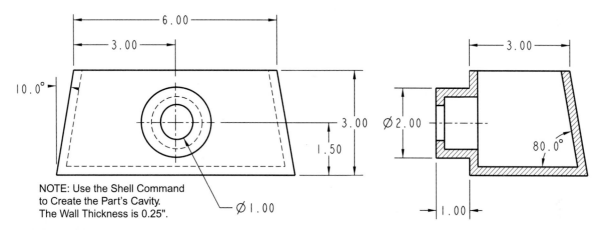

Figure 3-80   Problem two

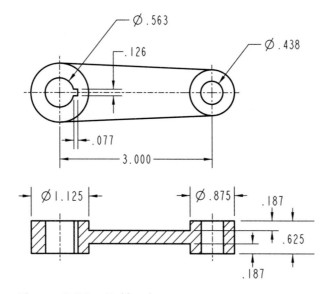

Figure 3-81   Problem three

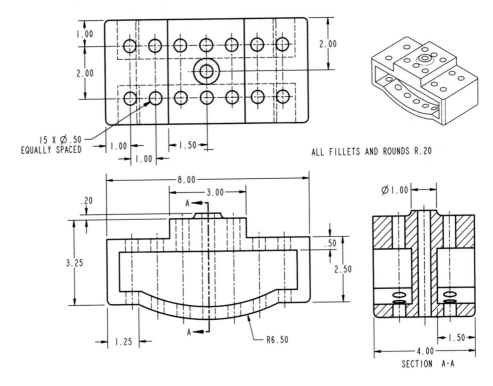

Figure 3-82   Problem four

QUESTIONS AND DISCUSSION

1. Describe the four different methods of placing a hole.

2. What is the difference between a Straight Hole, a Sketched Hole, and a Standard Hole? What are some uses of a Sketched Hole?

3. Describe methods for avoiding Round regeneration conflicts.

4. Describe the difference between a Simple Round and an Advanced Round.

5. What is the purpose of a Neutral Plane?

6. Compare and contrast the Protrusion >> Extrude option with the Rib command.

7. Describe the assumptions associated with the following pattern categories:

    a. Identical

    b. Varying

    c. General

8. In regard to the Pattern command, what is a leader dimension?

4

REVOLVED FEATURES

Introduction

Revolved feature construction techniques are common within Pro/ENGINEER. The most obvious revolved technique is the Revolve option found under the Protrusion and Cut commands. Other revolved feature construction techniques can be found with the sketched Hole option and the Shaft, Flange, and Neck commands. Upon finishing this chapter, you will be able to:

- Construct a sketched hole.
- Pattern a radial hole.
- Construct a revolved protrusion.
- Construct a revolved cut.

DEFINITIONS

Axis of revolution   The axis around which a section is revolved. Within Pro/ENGINEER, revolved features require a user-sketched centerline. This centerline serves as the Axis of Revolution.

Through >> Axis datum plane   A datum plane constructed through an axis.

REVOLVED FEATURE FUNDAMENTALS

A revolved feature is a section that is rotated around a centerline. For any type of revolved feature, within the sketcher environment, the user sketches the profile of the section to be revolved and the centerline to revolve about (Figure 4–1). Revolved features may be positive or negative space. A Revolved Protrusion is an example of a positive space feature. A Revolved Protrusion's negative space counterpart is the Revolved Cut. The Flange command is another example of a revolved positive space feature. Its counterpart negative space feature is the Neck command.

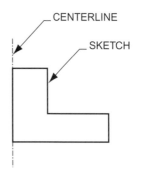

Figure 4-1 Revolved sketch

SKETCHING AND DIMENSIONING

As shown in Figure 4–1, the geometry of a revolved feature must be sketched on one side of the centerline and the section must be closed. One centerline must be sketched. If multiple centerlines exist in a section, the first one sketched will serve as the **axis of revolution.** Entities that lie on the axis of revolution will not serve as a replacement for the centerline.

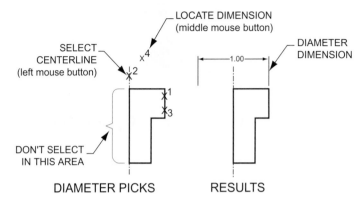

Figure 4-2   Diameter dimensions

Revolved features are often used to create cylindrical objects such as shafts and holes. Drafting standards require cylindrical objects to be dimensioned with a diameter value. This creates a unique situation within the sketcher environment. As shown in Figure 4–2, to dimension a revolved feature with a diameter value, perform the following steps:

1. Pick the geometry defining the outside edge of the feature.
2. Pick the centerline to serve as the axis of revolution.
3. Pick the geometry defining the outside edge of the feature.
4. Pick a location for the placement of the dimension text.

The resulting dimension should appear as shown in Figure 4–2.

> **MODELING POINT**   If a diameter dimension is not created, geometry was probably inadvertently selected instead of the required centerline. Select the centerline on the work screen where it is clear that a centerline is the only entity residing (Figure 4–2).

REVOLVED PROTRUSIONS AND CUTS

A Revolve option is available under the Protrusion and Cut commands. Revolved Protrusions are used to create positive space features while Revolved Cuts are used to create negative space features.

REVOLVED FEATURE PARAMETERS

As with extruded Protrusions and Cuts, a variety of options is available for defining revolved feature parameters.

REVOLVE DIRECTION

The Revolve direction attribute is similar to the Extrude direction attribute. Options are available for selecting a One Side or a Both Sides revolution. The One Side option will revolve the section from the sketching plane in one direction, while the Both Sides attribute will revolve the section both directions from the sketching plane.

ANGLE OF REVOLUTION

The Angle of Revolution parameter is similar to the Extrude option's depth parameter. This option is used to specify the number of degrees that the

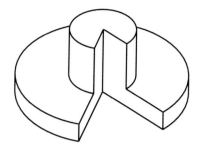

Figure 4-3   Variable angle of revolution

section will be revolved about the axis of revolution. The following options are available.

- **Variable** The Variable option is used to specify an angle of revolution less than 360 degrees. The angular parameter specified is modifiable (Figure 4–3).
- **90** The 90 option is used to rotate a section at an angle of 90 degrees.
- **180** The 180 option is used to rotate a section at an angle of 180 degrees.
- **270** The 270 option is used to rotate a section at an angle of 270 degrees.
- **360** The 360 option is used to rotate a section a full 360 degrees.
- **UpToPnt/Vtx** The UpToPnt/Vtx option is used to revolve a section up to a selected point or vertex.
- **UpTo Plane** The UpTo Plane option is used to revolve a section up to a selected plane.

CREATING A REVOLVED PROTRUSION

The Revolve option is used extensively for creating base geometric features. The following is a step-by-step guide for creating a revolved Protrusion.

STEP 1: Select FEATURE >> CREATE >> PROTRUSION.

STEP 2: Select REVOLVE on the Solid Options menu (Figure 4–4).

STEP 3: Select between a SOLID and THIN feature, then select DONE.

Just like extruded features, revolved features can be created as a solid or a thin feature. Solid features have mass while thin features have a user-defined wall thickness.

STEP 4: Select either ONE SIDE or BOTH SIDES, then select DONE.

STEP 5: Select a sketching plane and orient the sketcher environment.

Revolved features may be sketched on a part plane or datum plane. Additionally, an on-the-fly datum plane can be created with the Make Datum option.

STEP 6: Use the CENTERLINE icon to sketch a centerline (Figure 4–5).

The Centerline icon is located behind the Line icon. A sketched centerline is a requirement for any revolved feature. For Revolved Protrusions and Cuts, the centerline can be sketched at any orientation. For the first geometric feature of a part and when sketching on Pro/ENGINEER's default datum planes, it is recommended that you sketch the centerline of a revolved feature aligned with the edge of a datum plane. Consider design intent when sketching the centerline.

Figure 4-4   Revolve option

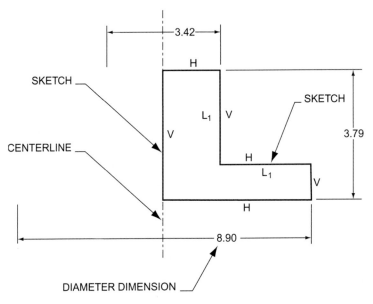

Figure 4-5   Section sketching

MODELING POINT   A centerline does not have to be the first entity sketched for a revolved feature. If a centerline is not present when selecting Done to exit the sketcher environment, Pro/ENGINEER will provide a warning message.

STEP   7:   **Sketch the Section (Figure 4–5).**

Use appropriate sketching tools to create the section. A revolved section must be closed and must lie completely on one side of the centerline.

STEP   8:   ✔ **Select the Continue icon to exit the sketcher environment.**

STEP   9:   **Specify an angle of revolution.**

STEP 10:   **Preview the feature on the Feature Definition dialog box, then select OKAY to create the feature (Figure 4–6).**

The Preview option is located on the Feature Definition dialog box. If the feature is not defined correctly, use the Define option to make changes to an element.

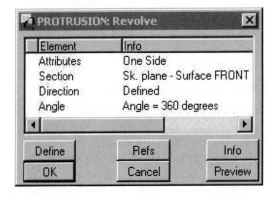

Figure 4-6   Revolve Feature Definition dialog box

REVOLVED HOLE OPTIONS

Pro/ENGINEER provides three hole types: straight, sketched, and standard. Straight holes have a constant diameter throughout the length of the feature. Sketched holes are used to create a unique profile, such as exist with counterbored and countersunk holes. In addition to the three types of holes, Pro/ENGINEER provides five placement options: Linear, Coaxial, Radial, Diameter, and On Point. The Linear option is used to locate a hole from two reference edges, while the Coaxial option is used to locate a hole's centerline coincident with an existing axis. The Radial option is used to locate a hole at a distance from an axis and at an angle to a reference plane (Figure 4–7). The hole's distance from the reference axis is defined by a radius value. Like the Radial option, the Diameter option is used to locate a hole at a distance from an axis and at an angle to a reference plane (Figure 4–8). With the Diameter option, the hole's distance from the reference axis is defined by a diameter value.

SKETCHED HOLES

The Sketched Hole option requires the user to sketch the profile of the hole (Figure 4–9). Most normal sketching tools can be used. The hole sketcher environment does not provide an option for specifying references, though. A sketched hole is created originally independent of any specific part features and later placed according to the hole placement option being used (e.g., Linear, Coaxial, Diameter, Radial, or On Point).

When sketching a hole, a vertical, user-created centerline is required. Within the sketcher environment, use the Centerline icon to create this entity. All sketched entities must be created on one side of the centerline and must be closed. A geometric entity may be placed on top of the centerline, but the centerline cannot serve as an element of the hole profile. Additionally, one sketched entity must lie perpendicular to the centerline. This entity will be aligned with the placement plane when placing the hole. For sketched holes with multiple perpendicular lines, the uppermost line within the sketcher environment will serve as the placement reference.

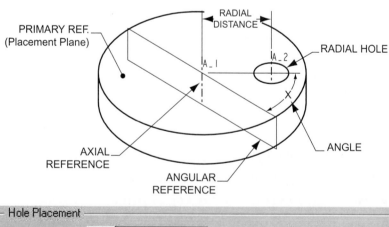

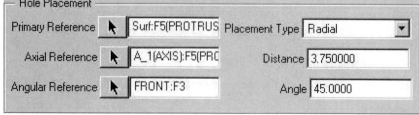

Figure 4-7   Radial hole placement

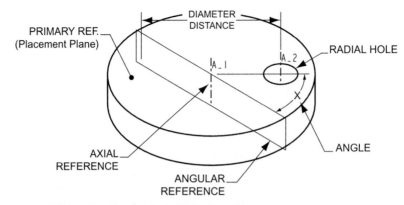

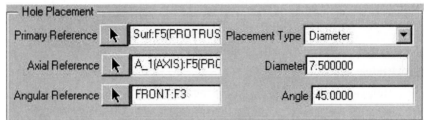

Figure 4-8   Diameter hole placement

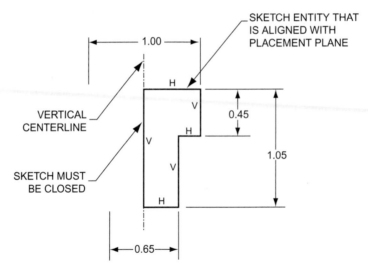

Figure 4-9   Sketching a counterbored hole

CREATING A SKETCHED HOLE

Perform the following steps to create a sketched hole.

STEP 1: **Select FEATURE >> CREATE >> HOLE.**

STEP 2: **Select SKETCHED as the Hole Type.**

Once the Sketched option is selected, Pro/ENGINEER will launch a sketcher environment for the creation of the hole's profile.

STEP 3: **Within the sketcher environment, use the CENTERLINE icon to create a vertical centerline (Figure 4–9).**

All revolved features require a centerline within the sketch. For a sketched hole, this centerline must be vertical.

STEP 4: Sketch the profile of the hole (Figure 4–9).

Sketch entities of the hole must be created on one side of the centerline and must be completely closed. The centerline will not serve as an entity of the sketch.

One sketched entity must be created perpendicular to the centerline. This entity will be used to align the hole with the placement plane. For multiple perpendicular entities, the uppermost one in the sketcher environment will serve this purpose.

STEP 5: ⟷ **Dimension the sketch to meet design intent.**

Use the Dimension option to dimension the sketch to meet design intent. Since holes are defined with diameter values, use the diameter dimensioning technique described previously in this chapter.

STEP 6: Modify dimension values.

STEP 7: ✔ **Select the Continue icon to exit the sketcher environment.**

STEP 8: Dependent upon the hole's placement option, select the hole's Primary Reference.

Place the hole according to requirements for a Linear, Coaxial, Radial, Diameter, or On Point hole. For the following are the primary references for each placement type:

- **Linear** Placement plane.
- **Coaxial** Axial reference.
- **Radial** Placement plane.
- **Diameter** Placement plane.
- **On Point** Datum point reference.

STEP 9: Select the hole's placement type.

Hole placement types available include: Linear, Coaxial, Radial, Diameter, and On Point.

STEP 10: Select the remaining placement references as required.

STEP 11: ✔ **Select the Built Feature icon on the Hole dialog box.**

RADIAL HOLE PLACEMENTS

The Radial and Diameter hole placement options are used frequently with the Pattern command to create a radial pattern of a hole (Figure 4–10). The Radial and Diameter options place a hole at a user-specified distance from an existing axis and at an angle to a reference

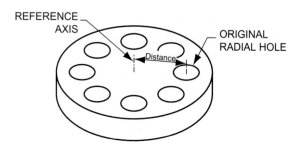

Figure 4-10 Patterned hole

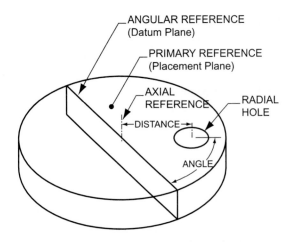

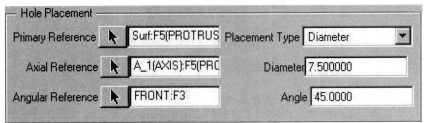

Figure 4–11   Radial hole creation

plane (Figure 4–11). With patterned holes, this angular dimension is used as the leader dimension for creating the pattern.

CREATING A STRAIGHT-DIAMETER (RADIAL) HOLE

Perform the following steps to place a Diameter Hole (Figure 4–11):

STEP **1:** **Select FEATURE >> CREATE >> HOLE.**

STEP **2:** **On the Hole dialog box, select STRAIGHT as the Hole Type.**

Radial holes may be created with the Sketched and Standard hole types also.

STEP **3:** **Enter the hole's Diameter value.**

STEP **4:** **Enter the hole's Depth One parameter (and value if required).**

STEP **5:** **Pick the hole's Placement Point (Primary Reference).**

STEP **6:** **Select DIAMETER (or Radial) as the hole's Placement Type (Figure 4–11).**

Other options available include Linear and Coaxial. The Linear option will locate a hole from two reference edges, while the Coaxial option will place the hole's centerline coincident with an existing axis.

STEP **7:** **Select the Axial Reference for the hole.**

The hole will be located at a user-specified distance from the selected axis.

STEP **8:** **Enter the Diameter distance value for the Radial Hole.**

In the Diameter textbox, enter the distance value for the Diameter hole. The diameter value will be twice a corresponding radial value. As an example, if the hole is to be located three inches from the axial reference, then you should enter six as the diameter value. This option is often used to create a Bolt-Circle hole pattern.

STEP **9:** **Select a Reference Plane, then enter an angular value from the plane (see Figure 4–11).**

Select an existing planar surface then enter an angular value. The hole will be located from the reference plane at the specified angle. A datum plane or planar surface may be selected.

STEP 10: **Select the Build Feature checkmark.**

ROTATIONAL PATTERNS

The Pattern command can be used to create a radial pattern of a Hole feature or to create a rotational copy of a sketched feature. To create a rotational pattern, an angular dimension must exist that defines the feature to be patterned. There are two common situations when a rotational pattern can be used.

RADIAL PATTERN

The Pattern command can be used to create an angular copy of a radial hole (Figure 4–12). The Hole-Radial (or Diameter) option places a feature by entering a distance from a selected axis and by entering an angle to a reference plane. This angle can be used as the leader dimension in the pattern creation process.

ROTATIONAL PATTERN

The Pattern option can be used to create a rotational copy of a sketched feature (Figure 4–13). The feature's section has to reference a **Through Axis datum plane** created with the Through and Angle constraint options, or the feature has to be sketched on a Through Axis datum plane. The angular dimension defining the sketching plane is used to pattern the feature.

INSTRUCTIONAL NOTE For more information on Patterns, see Chapter 3.

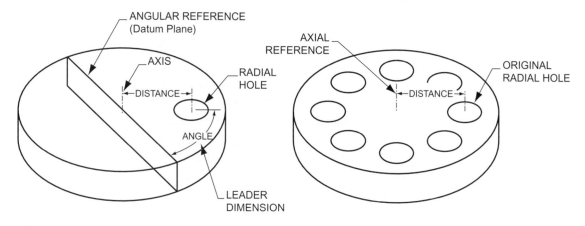

Figure 4-12 Radial pattern

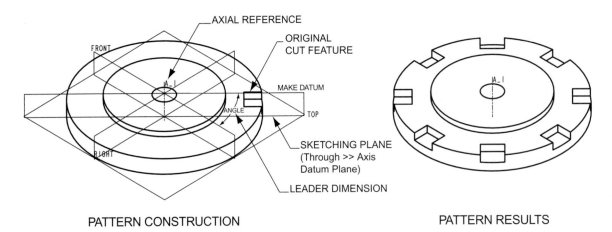

PATTERN CONSTRUCTION   PATTERN RESULTS

Figure 4-13   Rotational pattern

CREATING A RADIAL PATTERN

The following is a step-by-step approach for creating a unidirectional radial pattern. Two directional patterns can be used also. Additionally, dimensions other than the leader dimension can be varied.

STEP 1:   Select FEATURE >> PATTERN.

STEP 2:   Select the feature to pattern.

On the work screen or on the Model Tree, select a hole or sketched feature. The feature has to include an angular dimension that will serve as the leader dimension.

STEP 3:   Select a Pattern Option on the Pattern Options menu.

On the Pattern Options menu, select Identical, Varying, or General. Identical patterns require the most assumptions. With this option, instances of a pattern cannot intersect other instances or the edge of the placement plane. Additionally, an Identical pattern must exist on one placement plane only. The General option does not require any assumptions, but takes longer to regenerate.

STEP 4:   Select VALUE on the Pattern Dimension Increment menu.

STEP 5:   Select an angular leader dimension for use in varying the feature (see Figures 4–12 and 4–13).

To create a rotational pattern of a feature, the feature must have an angular dimension. Holes placed with the Radial placement type incorporate a reference angle dimension that locates the hole from a reference plane. Features sketched on an on-the-fly datum plane can be patterned if the datum plane is defined with an Angle constraint.

STEP 6:   Enter a dimension increment value.

Enter the number of degrees that the feature will be incremented. This value determines the spacing between each feature.

STEP 7:   Select DONE on the Exit menu.

Selecting Done will end the selection of varying dimensions for the first direction. Within a rotational pattern, dimensions other than the leader dimension can be varied also.

STEP 8:   Enter the number of instances in the first direction.

MODELING POINT   To create a bidirectional pattern, repeat steps 4–8. As with linear patterns, rotational patterns can be created in two directions.

STEP 9:   Select DONE to create the pattern.

DATUM AXES

Datum axes are used as references for the creation of features. As an example, they can be used to place a coaxial hole or a radial hole. Additionally, they often are used to create datum planes. When holes, cylinders, and revolve features are created, datum axes are created automatically. Datum axes created separately from a part feature are considered features. They are named in sequential order on the model tree starting with A_1.

CREATING DATUM AXES

Datum axes can be placed on parts or assemblies. Perform the following steps to create a datum axis in Part mode.

STEP 1:   On the Datum Toolbar, select the CREATE DATUM AXIS icon.

STEP 2:   Select a CONSTRAINT OPTION on the Datum Axis menu, then pick geometry appropriate to the selected constraint option.

Datum axes are created utilizing existing part features. Unlike datum planes, all datum axis constraint options are stand-alone. The following is a list of available constraint options.

THRU EDGE

The Through Edge option creates a datum axis through an existing edge of a part. The user has to select a geometric edge formed by two part surfaces.

NORMAL PLN

The Normal to Plane option creates a datum axis perpendicular to an existing plane. This plane can be a part or datum plane. Linear dimensional references are required.

PNT NORM PLN

This constraint option places a datum axis perpendicular to a plane and through an existing datum point. The datum point does not have to lie on the selected plane.

THRU CYL

Through Cylinder places a datum axis at the center of an existing cylinder. The surface can be a complete cylinder or a partial cylinder, such as a round.

TWO PLANES

The Two Planes option places a datum axis at the intersection of two planar surfaces or two datum planes. A common use of this feature is to place an axis at the intersection of two of Pro/ENGINEER's default orthogonal datum planes.

TWO PNT/VTX

This constraint option places a datum axis between two datum points, between two vertexes, or between a datum point and a vertex.

Pnt on Surf

This constraint option places a datum axis perpendicular to a surface and through a point that lies on the surface.

Tan Curve

Tangent Curve places a datum axis tangent to a curve or edge at a selected point. The point has to exist prior to the use of this option.

> **MODELING POINT**   Many constraint options require you to select multiple geometric features. Watch the message area carefully to know what object to select.

Step 3:   Select DONE.

The datum axis will be created when you select Done.

Summary

Some of the most common features created within Pro/ENGINEER are revolved features. Options that utilize a form of a revolved feature include Revolved Protrusions, Revolved Cuts, and Sketched Holes. One requirement of any revolved feature is the sketching of a centerline. Within the sketcher environment for a feature, entities of the sketch must lie completely on one side of the centerline.

REVOLVED FEATURES TUTORIAL

This tutorial exercise provides step-by-step instruction on how to model the part shown in Figure 4–14.

This tutorial will cover

- Creating a revolved protrusion.
- Creating a radial sketched hole.
- Creating a hole radial pattern.
- Creating a revolved Cut.
- Modifying the number of holes in a pattern.

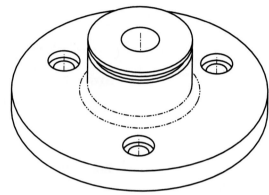

Figure 4-14   Finished part

CREATING A REVOLVED PROTRUSION

The first section of the Revolved tutorial will create the base revolved Protrusion shown in Figure 4–15.

STEP 1: **Start Pro/ENGINEER.**

STEP 2: **Establish an appropriate Working Directory.**

STEP 3: **Start a New part file named *revolve1* (use the default template file).**

Use either the File >> New option or the New icon to start a new part file named *revolved1*.

STEP 4: **Create a Revolved Protrusion with the following options:**

- Protrusion command.
- Solid Revolve.
- One Sided.
- Datum Plane FRONT as the sketching plane.
- Default sketch plane orientation.

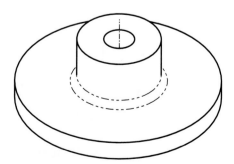

Figure 4-15   Revolved protrusion

Revolved positive space features are created under the Protrusion command. Revolved negative space features can be created under the Cut command.

STEP   5:   Close the References dialog box.

STEP   6:   Use CENTERLINE option to sketch a vertical centerline aligned with datum plane RIGHT (Figure 4–16).

The Centerline icon can be found under the Line icon. Revolved features require a user-created centerline in the sketching environment. The sketched section will be revolved around this centerline. With Revolved Protrusions and Cuts, this centerline can be created at any angle.

STEP   7:   Use the LINE option to sketch the section shown in Figure 4–17.

Sketch the section to the right of the centerline. Sketch all lines either horizontal or vertical as shown in the figure. Align the bottom of the sketch with datum plane TOP.

STEP   8:   Use the CIRCULAR FILLET option to create the fillet shown in Figure 4–17.

When using the Circular Fillet option, select the two entities to fillet between.

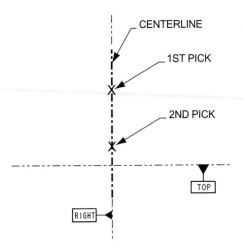

Figure 4-16   Sketching a centerline

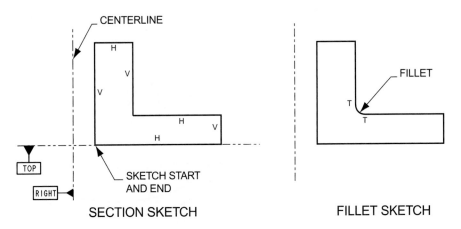

Figure 4-17   Sketching the section

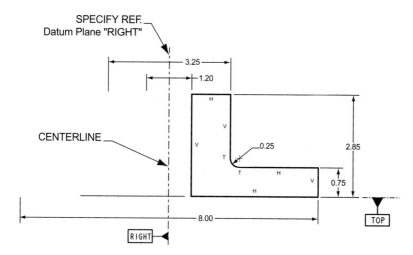

Figure 4-18   Dimensioning scheme

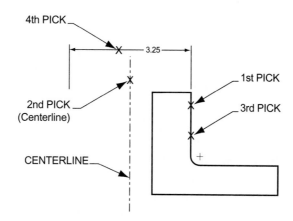

Figure 4-19   Diameter dimension

STEP   9:   Apply the dimensioning scheme shown in Figure 4–18. (Do Not Modify the Dimension Values within this Step.)

Use the Dimension option to create the dimensioning scheme shown in Figure 4–18. Pro/ENGINEER does not know the design intent for a feature. Due to this, the dimensions created by Intent Manager may not match those necessary for the design.

To create the diameter dimensions, perform the following four selections (Figure 4–19):

1.   Pick the outside edge of the entity (left mouse Button)
2.   Pick the centerline (left mouse Button)
3.   Pick the outside edge of the entity (left mouse Button)
4.   Place the dimension (middle mouse Button)

STEP 10:   Select the Modify option.

STEP 11:   On the work screen, select the dimension defining the height of the flange (the 2.85 dimension in Figure 4–18).

STEP 12:   On the work screen, select the remaining five dimensions.

After selecting the six dimensions defining the section, your Modify Dimension dialog box should appear similar to Figure 4–20.

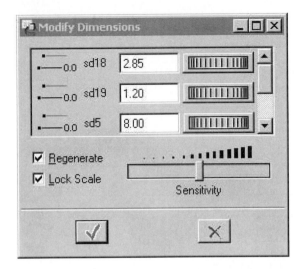

Figure 4-20   Modify Dimensions dialog box

STEP 13:   On the Modify Dimension dialog box, modify the flange's height dimension to have a value of 2.85 (select the tab key after entering value).

If you get unexpected results from your dimension modification, use Undo to correct any errors.

STEP 14:   Uncheck the LOCK SCALE option.

STEP 15:   On the dialog box, modify the remaining dimension values to match Figure 4–18.

STEP 16:   When all dimension values match the illustration, select the Regenerate checkmark on the dialog box.

STEP 17:   ✔ Select the Continue icon to exit the sketcher environment.

Do not select the Continue option until your section matches Figure 4–18.

STEP 18:   Specify a 360-degree revolution, then select DONE.

STEP 19:   Preview the feature, then select OKAY on the Feature Definition dialog box (Figure 4–21).

Your feature will look similar to Figure 4–22. The illustration shown displays the part with Tangent Edges set to Phantom and the default orientation set to Isometric. To change the Tangent Edge display and default orientation of

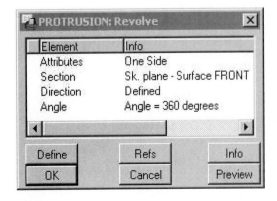

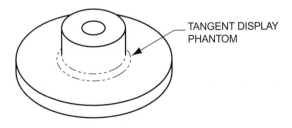

Figure 4-21   Revolve Feature Definition dialog box   **Figure 4-22**   Finished feature

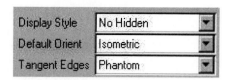

Figure 4-23   Environment options

a part, make the adjustments to the Environment dialog box (Utilities >> Environment), as shown in Figure 4–23.

CREATING A RADIAL SKETCHED HOLE

This segment of the tutorial will create the sketched Radial Hole shown in Figure 4–24.

STEP 1: **Select FEATURE >> CREATE >> HOLE.**

STEP 2: **Select SKETCHED as the hole type.**

The Sketched option will require you to sketch the hole's cross-section. After selecting this option, Pro/ENGINEER will open a sketcher environment.

STEP 3: 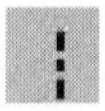   **Use the CENTERLINE option to create the vertical centerline shown in Figure 4–25.**

A vertical centerline is a requirement for a sketched Hole feature. The section will be revolved around the centerline.

STEP 4:    **Sketch the section shown in Figure 4–26.**

Use the Line option to sketch the section. For this specific section, sketch each entity either horizontally or vertically.

Figure 4-24   Finished hole feature

Figure 4-25   Sketching the centerline

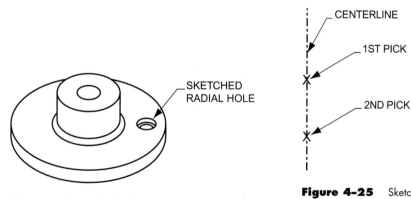

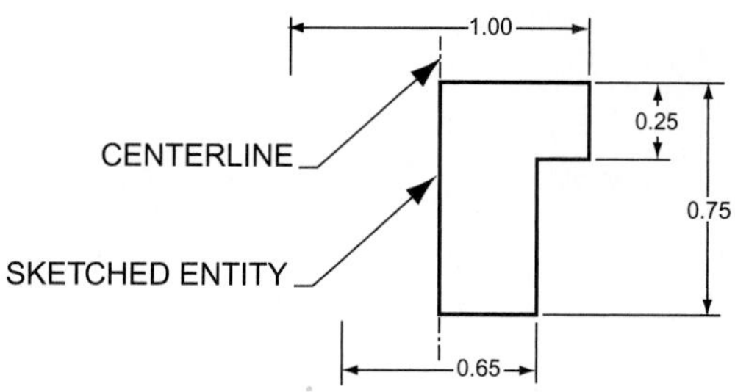

Figure 4-26   Sketched section

Sections of sketched holes must be completely enclosed. As shown in Figure 4–26, the section will be sketched on top of the previously created centerline. **The centerline will not serve as an entity to close the section.** You will receive an error message when exiting the sketcher environment if the section is not closed.

STEP 5: ⟷ **Apply the dimensioning scheme as shown in Figure 4–26.**

Use the Dimension option to match the dimensioning scheme shown in Figure 4–26. Holes are dimensioned with diameter values.

STEP 6: 📝 **Modify dimension values to match Figure 4–26.**

STEP 7: ✔ **Select the Continue icon when the section is correct.**

STEP 8: **Pick the hole's placement plane (Primary Reference) as shown in Figure 4–27.**

STEP 9: **Select DIAMETER as the hole's Placement Type (Figure 4–27).**

Using the Diameter option, the hole will be located at a distance (3 inches) from datum axis A_1 and at an angle to datum plane FRONT (45 degrees).

MODELING POINT   Creating a radial or diameter hole is the first step in modeling a rotationally patterned hole. The angular dimension created with a radial hole placement option is used as the leader dimension for patterning the hole around the center axis.

STEP 10: **Select the hole's AXIAL REFERENCE.**

As shown in Figure 4–27, select the reference axis. The hole will be located at a specified distance from the selected axis.

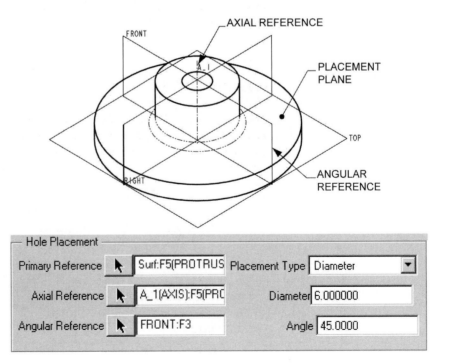

Figure 4-27   Hole placement references

STEP 11: **Enter 6.00 as the Diameter value.**

The Diameter value is double the distance from the Axial Reference to the center of the hole.

STEP 12: **Select datum plane FRONT as the hole's ANGULAR REFERENCE plane (Figure 4–27).**

The hole will be placed at an angle to the reference plane. As shown in Figure 4–27, select datum plane FRONT as the reference plane.

STEP 13: **Enter 45 as the reference angle.**

STEP 14: Preview the feature.

STEP 15: Select the Build Feature checkmark.

STEP 16: **Save your part.**

CREATING A RADIAL HOLE PATTERN

This segment of the tutorial will create a radial pattern of the hole created in the previous section of this tutorial. The 45-degree angular reference dimension used to locate the hole will be used as the leader dimension within the pattern. The finished pattern is shown in Figure 4–28.

STEP 1: **Select FEATURE >> PATTERN.**

> **MODELING POINT**   The Pattern command will create a rotational or linear pattern of a single feature. To create a pattern of multiple features, each feature must be grouped using the Group command. Within the Group menu is a separate Pattern option.

STEP 2: **On the Model Tree, select the previously created hole feature (Figure 4–29).**

STEP 3: **Select IDENTICAL >> DONE as the Pattern option.**

The Identical option allows several assumptions in the pattern creation process. Features patterned with this option must lie on the same placement plane, instances cannot intersect the placement plane's edge, and instances

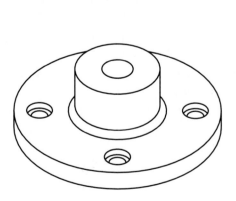

Figure 4-28   Finished pattern

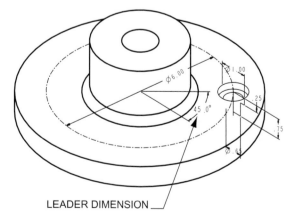

LEADER DIMENSION

Figure 4-29   Feature parameters

cannot intersect other instances. To allow instances to be placed on different planes and to allow instances to intersect an edge, use either Varying or General. To allow instances to intersect, use the General option.

STEP 4: As shown in Figure 4–29, pick the 45-degree angular dimension used to place the sketched radial hole.

The 45-degree angular dimension will be used as the leader dimension within the patterning process. This dimension will be varied to create the pattern.

STEP 5: Enter 90 as the increment value.

The leader dimension will be varied 90 degrees per instance of the hole feature. In other words, each hole within the pattern will be 90 degrees apart.

STEP 6: Select DONE on the Exit menu.

Selecting Done will end the varying process in the first direction of the pattern. Patterns may be unidirectional or bidirectional. For both cases, dimensions in addition to the leader dimension can be selected for varying. Selecting Done will exit the first direction of patterning.

STEP 7: Enter 4 as the number of instances in the first direction.

STEP 8: Select DONE on the Exit menu.

No instances of the hole will be created in the second direction. Selecting Done will create the pattern (Figure 4–28).

STEP 9: Save the part.

CREATING A REVOLVED CUT

This segment of the tutorial will create the Revolved Cut feature shown in Figure 4–30.

STEP   1: Select FEATURE >> CREATE >> CUT.

STEP   2: Select REVOLVE >> SOLID >> DONE.

STEP   3: Select ONE SIDE >> DONE.

The One Side option will revolve the section one direction from the sketching plane.

STEP   4: Select datum plane FRONT as the sketching plane (Figure 4–31), then select OKAYAY on the Direction menu.

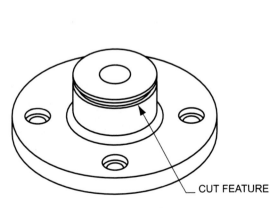

Figure 4-30   Revolved Cut feature

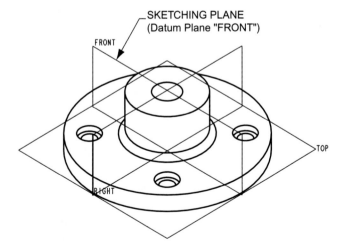

Figure 4-31   Sketching plane

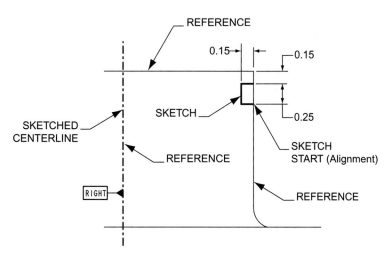

Figure 4–32   Section creation

STEP 5: **Select DEFAULT to set the sketcher environment's orientation.**

Optionally, you could select the Top option, then pick the top of the flange. This would orient the flange toward the top of the sketcher environment.

STEP 6: **Within the sketcher environment, specify the three references shown in Figure 4–32.**

Use the References dialog box to specify the three references shown in the illustration.

STEP 7: **Sketch the section shown in Figure 4–32.**

Sketch the section as shown. To better view the entities being sketched, use a zoom option to match the zoom extents shown in Figure 4–32. By aligning with the outside edge of the existing flange, a closed section is not necessary. After sketching the section, apply the dimensioning scheme shown, then modify the dimension values accordingly.

STEP 8: **Create a Centerline aligned with datum plane RIGHT (see Figure 4–32).**

Sketch a vertical centerline aligned with datum plane RIGHT. This centerline will serve as the required axis of revolution.

STEP 9: **Select the Continue icon to exit the sketcher environment.**

STEP 10: **Select the Default Material Removal Side.**

STEP 11: **Select 360 >> DONE as the Revolve Angular value.**

STEP 12: **Preview the feature, then select OKAY on the Feature Definition dialog box.**

MODIFYING THE NUMBER OF HOLES

In this segment of the tutorial you will use the Modify command to change the number of holes around the bolt-circle. The final pattern will appear as shown in Figure 4–33.

STEP 1: **Select MODIFY on Pro/ENGINEER's Menu Manager.**

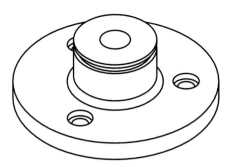

Figure 4–33   Finished part

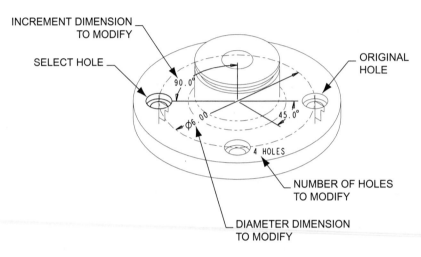

Figure 4–34   Pattern parameters

STEP 2: As shown in Figure 4–34, select one of the three patterned instances of the original sketched hole.

Select one of the patterned instances of the hole, not the original. As shown in Figure 4–34, an instance of a rotated pattern will show the original leader dimension value (45 degrees for this part) and the increment value (90 degrees). Also shown is the number of instances of the patterned hole (4 holes).

STEP 3: Select the 90-degree value and modify it to equal 120.

STEP 4: Select the 4 HOLES parameter and modify it to equal 3.

When modifying the number of instances of a pattern, you have to pick the actual text defining the value. In this case, select the number 4.

STEP 5: Select the 6.00 diameter dimension and modify it to equal 5.50.

STEP 6: REGENERATE the Part.

Your part should appear similar to Figure 4–33.

STEP 7: Save your part.

STEP 8: Purge old versions of the part by using the FILE >> DELETE >> OLD VERSIONS option.

PROBLEMS

1. Using Pro/ENGINEER's Part mode, model the parts shown in Figures 4–35 and 4–36. For both parts, create the base protrusion as a revolved feature.

2. Use Pro/ENGINEER to model the parts shown in Figures 4–37, 4–38, and 4–39.

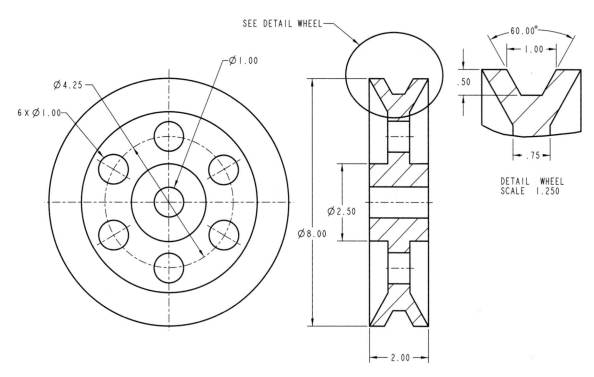

Figure 4-35   Problem one (a)

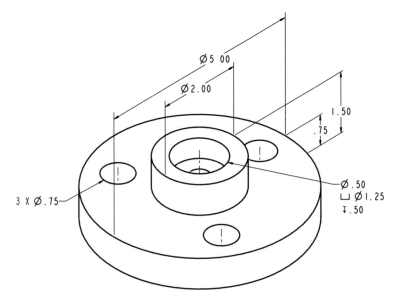

Figure 4-36   Problem one (b)

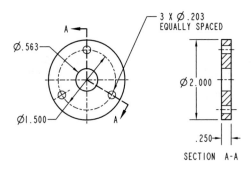

Figure 4-37   Problem two (a)

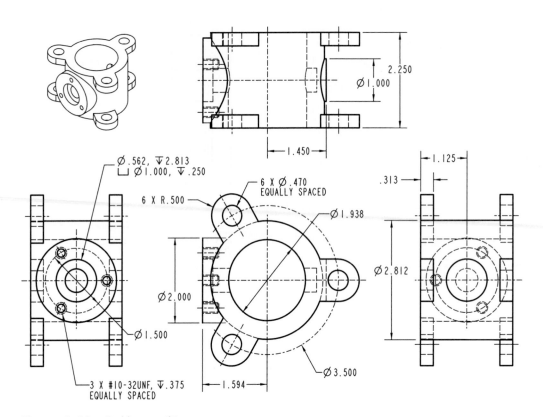

Figure 4-38   Problem two (b)

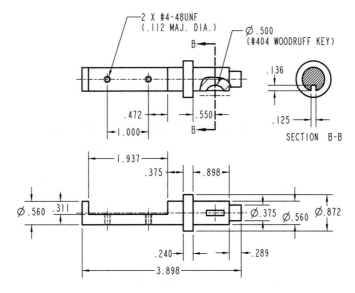

Figure 4-39 Problem two (c)

QUESTIONS AND DISCUSSION

1. List and describe various revolved features available within Pro/ENGINEER.

2. Describe the necessary requirements for sketching a revolved feature's section.

3. Describe the procedure for creating a diameter dimension within a section that will be revolved.

4. What revolved feature requires a vertical centerline?

5. Describe the difference between a hole placed with the Radial option and a hole placed with the Diameter option.

6. What dimension type that defines a feature must be available to allow for the creation of a rotational pattern?

5

FEATURE MANIPULATION TOOLS

Introduction

Commands such as Protrusion, Cut, Rib, Datum, and Hole are used to create part features. Pro/ENGINEER provides a variety of tools for manipulating existing features. As an example, multiple features can be combined with the Group option. When a group is created, it can be manipulated with options such as Copy and Pattern. Other manipulation tools covered in this chapter include User-Defined Features, Relations, and Family Tables. Upon finishing this chapter, you will be able to

- Combine features as a local group.
- Pattern a local group.
- Copy features in a linear direction.
- Mirror features across a plane.
- Copy-rotate features.
- Copy features by specifying new references.
- Add a dimension relationship to a part.

DEFINITIONS

Group   A collection of features combined to serve a common purpose.

Local group   A combination of features available within the current model.

Relation   An explicit relationship that exists between dimensions and/or parameters.

User defined feature   A feature that is stored to disk as a group and can be used in other models.

GROUPING FEATURES

Most Pro/ENGINEER modification options are utilized to manipulate individual features. Often, it is desirable to manipulate multiple features together. As an example, Figure 5–1 shows a part with a rotationally patterned boss, round, and hole. The normal Pattern command is used on individual features, not multiple features. The grouped boss, round, and hole features shown in the illustration were patterned using the Group >> Pattern option.

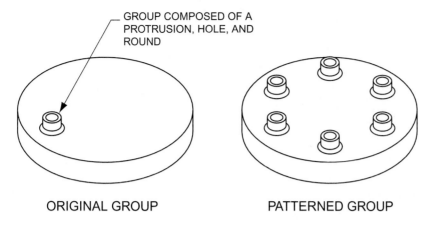

ORIGINAL GROUP   PATTERNED GROUP

Figure 5-1   Patterned group

THE GROUP MENU

The Group menu provides options for creating and manipulating groups. The following options exist:

CREATE

The Create option is used to create a group. Two options are available. The first option is to place a User-Defined Feature. When a UDF is placed in an object, it becomes a grouped feature. The second option is to create a Local Group. Local Groups are available only in the current model.

PATTERN

The Pattern option is used to create a rotational or linear pattern of a group. The Group >> Pattern option works similarly to the Feature >> Pattern command.

REPLACE

The Replace option is used to replace a UDF that has been placed in an object. The new UDF must have the same number and type of references. A Local Group cannot be replaced.

UNPATTERN

Grouped features are combined on the Model Tree as essentially one feature. Groups patterned are grouped together on the Model Tree as a patterned feature. The Unpattern option will break the pattern relationship, creating individual groups of each pattern instance.

UNGROUP

Features combined with the Group option can be ungrouped with the Ungroup option.

MODELING POINT   The Group, Pattern, Unpattern, and Ungroup options can be used in sequence to copy multiple features in a way that will not require each feature to lose its individual identity. After a group has been patterned, it can be unpatterned with the Unpattern option. After performing an unpattern, each group that made up the pattern can be ungrouped with the Ungroup option.

GROUP TYPES

There are two types of groups: User-Defined Features and Local Groups. A **User-Defined Feature** (UDF) is a feature that has been grouped and saved to disk, often in a UDF library.

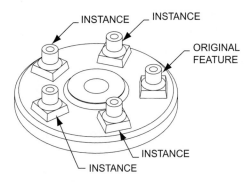

Figure 5-2   Pattern of features

A UDF can be retrieved and placed in the current working model or in another model. When a UDF is placed in an object, it becomes a grouped feature on the Model Tree.

A **Local Group** is a combination of features available within the current model only. The Group option allows multiple features to be grouped and patterned. Features combined to form a local group must be adjacent to each other in the order of regeneration. Due to this, it is important during the modeling process to place intended group features next to each other on the Model Tree. To create a Local Group, perform the following steps:

STEP 1:   **Select FEATURE >> GROUP >> CREATE.**

After selecting Create, Pro/ENGINEER will launch the Open dialog box.

STEP 2:   **Select CANCEL on the Open dialog box.**

The Group >> Create option defaults to placing a User-Defined Feature. Select the Cancel option to exit from the Open dialog box.

STEP 3:   **Select LOCAL GROUP on the Create Group menu, then enter a name for the group.**

STEP 4:   **On the Model Tree, select the features to include in the group.**

Features may be selected directly on the work screen or from the model tree. Features in a group must be adjacent to each other on the model tree. When selecting features on the model tree, the beginning and ending features of the group may be selected.

STEP 5:   **Select DONE on the Create Group menu.**

GROUP PATTERNS

A common reason to create groups is to pattern multiple features (Figure 5–2). Within the Group menu is a group Pattern option. Local Groups and User-Defined Features can be patterned similarly to how individual features can be patterned with the Feature >> Pattern command. Linear and Rotational patterns may be created. While the Feature >> Pattern command has an option for selecting between an Identical, Varying, and General option, the Group >> Pattern option accepts the Identical option by default.

COPYING FEATURES

While the Pattern command is used to create multiple instances of a feature in a rotational or linear fashion, the Copy command is used to make a single copy of a feature or features. While the Pattern command creates new instances of a feature by varying dimensions that define the feature, the Copy command creates a copy by changing the placement

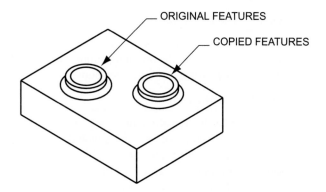

Figure 5-3   Copied features

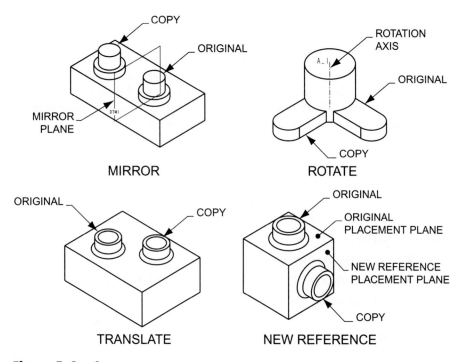

Figure 5-4   Copy options

of references and/or by changing dimension values. Figure 5–3 shows an example of a boss and coaxial hole that have been copied to another location. References required to define the location of the two features include a placement plane and two location edges. These references were changed to create the copy.

COPY OPTIONS

Unlike the Feature >> Pattern option, multiple features can be simultaneously copied with the Copy command. As shown in Figure 5–4, there are four basic types of copies that can be created.

MIRRORING FEATURES

The Mirror option is used to copy-mirror features about a plane. A mirrored image of the original feature is created.

ROTATIONAL COPIES

Rotate is a suboption under the Move option. Features can be copy-rotated around a datum curve, edge, axis, or coordinate system.

TRANSLATED FEATURES

Translate is a suboption under the Move option. Copied features are translated in a linear direction from the original features. The Same Refs option under the Copy menu creates a form of a translated copy.

NEW REFERENCES

Copies can be created of features by varying dimension values and by selecting new references. Examples of references include placement edges, reference axis, and placement plane. A feature can be copied with completely new references. The Same References option creates a copy of selected features, but does not allow the selection of alternate references.

INDEPENDENT VERSUS DEPENDENT

Features can be copied independent or dependent of their parent features. When a feature is copied as dependent, its parent features' dimension values govern its dimension values. If a dimension is changed in the parent feature, the corresponding dimension is changed in the copy. The Independent option allows copied features to be independent of parent features. When a feature is copied as independent, the copy's dimensions will remain independent of its parents'. The dimensions of the copy can be changed, and any changes to the parents' dimensions will not affect the child's dimensions.

MIRRORING FEATURES

The Mirror option creates a reflected image of selected features. To construct a mirror, select the features to be copied, and then select a mirroring plane. Perform the following steps to create a mirrored feature.

STEP 1: Select FEATURE >> COPY.

STEP 2: Select the MIRROR option.

STEP 3: Select either DEPENDENT or INDEPENDENT, then select DONE.

The Independent option will make the new copy's dimension values independent of any parent features.

STEP 4: Select features to be mirrored.

Features can be selected by picking them on the work screen or on the Model Tree. Additionally, the All Feat option can be used to mirror every available feature of a part. The Select option on the Copy menu can be used to select features during later steps.

STEP 5: Select DONE to finish selecting features.

STEP 6: Select a Mirroring Plane (see Figure 5–5).

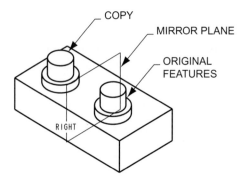

Figure 5-5   *Mirroring features*

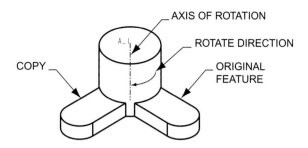

Figure 5-6   Rotating features

The mirroring plane can be any planar surface or datum plane. If a plane is not available, one can be created with the Make Datum option. The Mirror option does not utilize a Feature Definition dialog box. After the mirroring plane is selected, the copy is created.

ROTATING FEATURES

The Rotate option is located under the Move menu option. It is used to copy features by rotating them around an axis, edge, datum curve, or coordinate system. As with other Copy options, multiple features can be copied. Perform the following steps to copy-rotate selected features (see Figure 5–6):

STEP 1: **Select FEATURE >> COPY >> MOVE.**

STEP 2: **Select between INDEPENDENT and DEPENDENT.**

When features are copied as Dependent, their dimension values are dependent upon the original feature. Any changes made to the parent feature will be reflected in the copy.

STEP 3: **Select DONE on the Copy Feature menu.**

STEP 4: **Select features to copy, then select DONE.**

STEP 5: **Select ROTATE on the Move Feature menu (Figure 5–7).**

STEP 6: **Select either CRV/EDG/AXIS or CSYS on the Selection Direction menu, then select the appropriate entity on the work screen (Figure 5–8).**

Features can be copy-rotated around a datum curve, edge, axis, or coordinate system. After choosing a rotation type, select the appropriate entity on the work screen.

STEP 7: **Select OKAY for direction of rotation or FLIP to change direction.**

Pro/ENGINEER utilizes the right-hand rule to determine the direction of rotation. With the right-hand rule, point the thumb of your right hand in the direction of the arrow shown on the work screen. The remaining fingers of your right hand point in the direction of rotation.

STEP 8: **Enter the degrees of rotation.**

Enter the number of degrees that the selected features will be rotated.

STEP 9: **Select DONE MOVE on the Move Feature menu (Figure 5–7).**

STEP 10: **Select DONE on the Variable Dimension menu.**

Notice in Figure 5–9 the options for checking a dimension. The GP VAR DIMS menu allows for the selection and varying of dimension values during the copy process.

STEP 11: **Select OKAY on the Feature Definition dialog box.**

Figure 5-7   Move Feature menu

Figure 5-8   Selection Direction menu

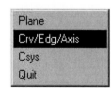

Figure 5-9   Variable Dimension menu

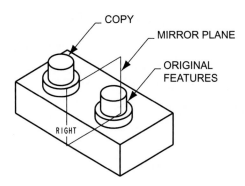

Figure 5-10   Translation

TRANSLATING FEATURES

Features can be copied in a linear direction using the Translate option. Features are copied perpendicular to a selected plane (see Figure 5–10).

STEP   1:   **Select FEATURE >> COPY >> MOVE.**

STEP   2:   **Select between INDEPENDENT and DEPENDENT.**

When features are copied as Dependent, their dimension values are dependent upon the original feature. Any changes made to the parent feature will be reflected in the copy.

STEP   3:   **Select DONE on the Copy Feature menu.**

STEP   4:   **Select features to copy, then select DONE.**

STEP   5:   **Select TRANSLATE on the Move Feature menu.**

STEP   6:   **Select PLANE on the GEN SEL DIR menu, then select a plane on the work screen.**

Select a planar surface or a datum plane. Features will be copied perpendicular to the selected plane. The Make Datum option is available to create an on-the-fly datum plane.

STEP   7:   **Accept or FLIP the Translate Direction.**

On the work screen, Pro/ENGINEER will graphically display the direction of translation.

STEP   8:   **Enter the translation value.**

Enter the value that the copied features will be offset from the original features.

STEP   9:   **Select DONE MOVE on the Move Feature menu.**

STEP 10:   **Select DONE on the Variable Dimension menu.**

As with the Rotate option, the GP VAR DIMS menu allows for the selection and varying of dimension values during the copy process.

STEP 11:   **Select OKAY on the Feature Definition dialog box.**

COPYING WITH NEW REFERENCES

The New Refs option copies selected features by specifying new references and by varying feature dimensions. Examples of references that can be changed include sketching planes and reference edges. Perform the following steps to copy a feature with new references (see Figure 5–11).

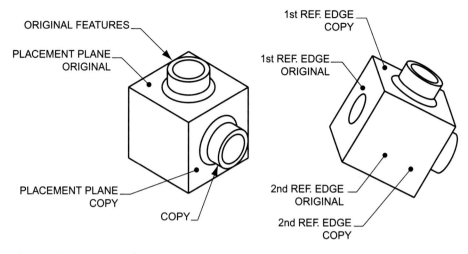

Figure 5-11   New Reference option

INSTRUCTIONAL NOTE   When Pro/ENGINEER requests the selection of a new reference for a copy (or other options such as Reroute), it highlights the old reference with the Section color setting. The Section color is set under the View >> Display Settings >> System Colors option. If the default Reference Color is hard to see, its setting should be changed.

Figure 5-12   Variable Dimension menu

STEP 1:   Select FEATURE >> COPY >> NEW REFS.

STEP 2:   Select between INDEPENDENT and DEPENDENT, then select DONE.

STEP 3:   Select features that will be copied, then select DONE.

STEP 4:   Select dimensions to vary (Optional Step).

Dimensions defining a feature can be varied during the copy process (Figure 5–12). Pro/ENGINEER will display the dimensions defining the selected features. On the work screen or on the GP VAR DIMS menu, select the dimensions to vary.

STEP 5:   Select DONE on the Variable Dimension menu (Figure 5–12).

STEP 6:   Choose an option for each highlighted reference, then perform the appropriate reference selection (Figure 5–13).

In sequential order, Pro/ENGINEER will highlight each reference in the established Reference (section) Color. For each reference, you must perform one of the following options:

- **Alternate**   This option requires the selection of a new reference. Select the appropriate new reference for the copy.

- **Same**   This option keeps the highlighted reference for the copy.

- **Skip**   This option allows you to skip the definition of a new reference for a copy. This reference must be defined later with the Redefine option.

- **Ref Info**   This option provides information about the reference.

Figure 5-13   Reference selection

STEP 7:   Select DONE on the Group Place menu.

RELATIONS

Mathematical and conditional relationships can be established between dimension values. Chapter 2 introduced the utilization of **relations** within the sketcher environment.

Table 5-1 Mathematical operations

Operator or Function	Meaning and Example
+	Addition d1 = d2 + d3
−	Subtraction d1 = d2 − d3
*	Multiplication d1 = d2 * d3
/	Division d1 = d2/d3
^	Exponentiation d1^2
( )	Group parentheses d1 = (d2 + d3)/d4
=	Equal to d1 = d2
cos ( )	cosine d1 = d2/cos(d3)
tan ( )	tangent d1 = d3 * tan (d4)
sin ( )	sine d2 = d3/sin (d2)
sqrt ( )	square root d1 = sqrt (d2) + sqrt (d3)
==	Equal to d1 == 5.0
>	Greater than d2 > d1
<	Less than d3 < d5
>=	Greater than or equal to d3 >= d4
<=	Less than or equal d5 <= d6
!=	Not equal to d1 != d2 * 5
|	Or (d1 * d2) | (d3 * d4)
&	And (d1 * d2) & (d3 * d4)
~	Not (d3 * d4) ~ (d5 * d6)

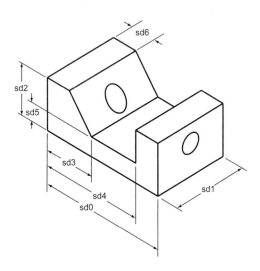

Figure 5-14   Dimension symbols

Relations within the sketcher environment are used to establish dimensional relationships between dimensions of a feature. The Relations command found under the Part menu is used to establish relationships between any two dimensions of a part. Within Assembly mode, the Relations option can be used to establish relationships between dimensions from different parts.

Dimensions can be shown with numeric values or as symbols. Figure 5–14 shows an example of a part with dimensions shown as symbols. Dimension values are displayed as a *d* followed by the dimension number (i.e., *d3*). Other parameters that can be displayed symbolically include Reference Dimensions (i.e., *rd3*), Plus-Minus Symmetrical Tolerance mode (i.e., *tpm4*), Positive Plus-Minus Tolerance mode (i.e., *tp4*), Negative Plus-Minus Tolerance mode (i.e., *tm4*), and number of instances of a feature (i.e., *p5*).

Most algebraic operators and functions can be used to define a relation. Additionally, most comparison operators can be used. Table 5–1 provides a list of mathematical operations, functions, and comparisons supported in relation statements. All trigonometric functions use degrees.

CONDITIONAL STATEMENTS

Pro/ENGINEER's Relations command has the ability to utilize conditional statements for the purpose of capturing design intent. As an example, Figure 5–15 shows a flange that incorporates holes on a bolt circle centerline. In a model such as this, the holes are created typically as a patterned radial hole. Imagine a situation where the number of holes and the diameter of the bolt circle are governed by the diameter of the flange. Suppose, in this example, that the bolt circle diameter is always 2 inches less than the diameter of the flange. Also suppose that the design requires 4 holes on the bolt circle if the diameter of the flange is 10 inches or less and 6 holes if the flange diameter is greater than 10 inches. Using the following relational statements, parameters can be established that control this design intent for the bolt circle pattern.

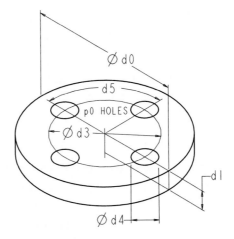

Figure 5-15   Dimension symbols

IF d0 <= 10	(Line 1)
p0 = 4	(Line 2)
d5 = 90	(Line 3)
ELSE	(Line 4)
p0 = 6	(Line 5)
d5 = 60	(Line 6)
ENDIF	(Line 7)
d3 = d0 $\times$ 2	(Line 8)

A conditional relation is defined between an IF statement and an ENDIF statement. The above example utilizes an IF-ELSE statement. With an IF-ELSE statement, if the condition is true, the expressions following the IF statement will occur. If the condition is not true, the condition following the ELSE statement will occur. The above example reads as follows:

> If the flange diameter is less than or equal to 10, the number of holes is equal to 4, and the angular spacing between each hole is 90 degrees; or else, the number of holes is equal to 6, and the angular spacing is equal to 60 degrees.

In the above example, the condition statement is the diameter of the flange (d0) being less than or equal to 10 (see line 1). If this is true, the number of holes on the bolt circle (p0) will be equal to 4, and the angle between each instance of the hole (d5) will be 90 degrees. If the flange diameter is not less than or equal to 10, the number of holes on the bolt circle will be 6, and the angle between instances will be 60 degrees. In a conditional statement, the ELSE expression shown in line 4 must occupy a line by itself. The expression ENDIF is used to end the conditional statement. Additionally, in the above example, line 8 is used to make the bolt circle diameter always 2 less than the flange diameter.

ADDING AND EDITING RELATIONS

Relations are added to an object using the Relations >> Add option. Dimension and parameter symbols are not automatically displayed upon selecting the Add option. Since these symbols are needed for the accurate creation of a relation, it is necessary to select each appropriate feature after selecting the Relations command. When a feature is selected in this manner, dimensions and parameters will be shown with their assigned dimension symbols.

Relations are added one relation at a time through a textbox. After typing a relation, use the Enter key to input the relation. For conditional statements, input each line one at a time, also through the textbox. Relations are evaluated within a model in the order in which they are defined. In most cases of conflict, the later relation overrides the former.

The Show Relations (Show Rel) and the Edit Relations (Edit Rel) options can be used to display defined relations. While the Show Rel option allows for the viewing of relations, the Edit Rel option allows for the modification of existing relations and for the creation of new relations.

Perform the following steps to add relations to a model.

STEP 1:   **Select RELATIONS on the Part menu.**

STEP 2:   **Select the Part or Feature applicable to the relation.**

Selecting a model will display the dimensions defining the selected part or feature. While in the Relations menu, dimension will be displayed in the symbol format. You can use the Switch Dim command from the Info menu to change the display of a dimension.

STEP 3:   **Select ADD on the Part menu.**

STEP 4:   **Type the appropriate relation in the textbox, then select the ENTER key.**

STEP 5:   **To end the adding of relations, select the ENTER key with a blank textbox.**

STEP 6:   **Regenerate the model.**

Defined relations will not take effect until the model is regenerated.

SUMMARY

The traditional way within Pro/ENGINEER to create features is to use commands such as Protrusion and Cut. Options exist that will optimize the modeling process by manipulating these features. Like most mid-range CAD packages, Pro/ENGINEER has commands for manipulating existing entities and features. Options are available, such as Pattern, that will allow for the copying and arraying of most features. Other options such as Copy also exist for creating instances of a feature. The understanding of these manipulation tools is critical for creating advanced parts within Pro/ENGINEER.

MANIPULATING TUTORIAL 1

This tutorial will create the part shown in Figure 5–16. The following topics will be covered:

- Creating a revolved protrusion.
- Creating an extruded protrusion.
- Creating a coaxial hole.
- Mirroring a feature.
- Rotating a feature.
- Adding dimensional relationships.

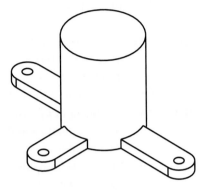

Figure 5-16   Finished part

CREATING THE BASE PROTRUSION

The first section of this tutorial will require you to create the Protrusion feature shown in Figure 5–17. Start Pro/ENGINEER and create a part model named *Rotate*.

Create the base protrusion of the part as a Revolve with the following requirements:

- Revolve the section shown in Figure 5–17.
- Use Pro/ENGINEER's default datum planes.
- Sketch the Revolved Protrusion on datum plane FRONT.
- Use a 360-degree revolution parameter.

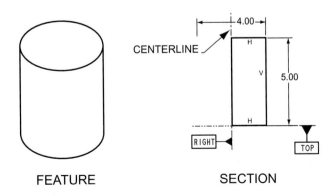

FEATURE   SECTION

Figure 5-17   Base protrusion and section

CREATING AN EXTRUDED PROTRUSION

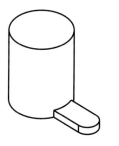

Figure 5-18　Extruded protrusion

This segment of the tutorial will create the Extruded Protrusion shown in Figure 5–18.

STEP 1: **Setup a Solid Extruded PROTRUSION for the new feature shown in Figure 5–18 by using the following options:**

- Use a One Side Extrude.
- Select datum plane TOP as the sketching plane.
- Use the Default sketching orientation.

STEP 2: **Once established in the sketcher environment, close the References dialog box.**

STEP 3: ☐ **Select the USE EDGE icon, then select the SINGLE option on the Type dialog box.**

The Use Edge option is used to project existing feature edges as sketch entities within the current sketching environment.

STEP 4: **As shown in Figure 5–19, select the outside edge of the base revolved feature in the two specified locations. (Only select each location once.)**

The Use Edge option will project the outside edge of the revolved feature onto the sketching plane as sketcher entities. The selected edges will become references within the sketching environment.

STEP 5: **Sketch the two line entities and the arc entity shown in Figure 5–20.**

Sketch the two new lines and the arc as shown. The ends of each line should be aligned with the edge of the existing revolved protrusion. Since the previous step of this tutorial turned the edges of the revolution into entities and into references, the lines will snap to the edge.

STEP 6: **Modify the dimension values to match Figure 5–20.**

STEP 7: ⌐ **Select the Trim Entities icon on the Sketcher toolbar.**

Note: ⌐ **The Trim Entities icon is located behind the Dynamic Trim icon.**

INSTRUCTIONAL NOTE　Ensure that you select the Trim Entities icon and not the Dynamic Trim icon. The Dynamic Trim option selects entities to trim by requiring the user to dynamically draw a construction spline.

With the Trim option, the two selected entities will be trimmed at their intersection point. You must pick the portion of each entity that should not be deleted.

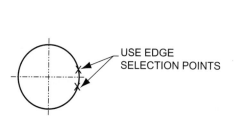

Figure 5-19　Selecting an edge

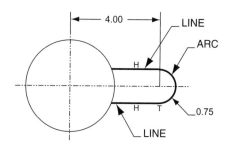

Figure 5-20　Sketched section

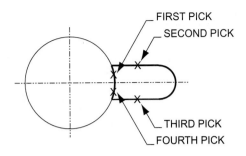

Figure 5–21   Trim selections

STEP 8: **Select the four locations shown in Figure 5–21.**

With the Trim option selected, the chosen locations identify the entities to trim and the portion of each entity to keep.

STEP 9: **When the section is complete, select the Continue icon.**

If you get an error message upon selecting the Continue option, you probably have section entities that overlap. Within any section, you cannot have sections that branch or sketch entities that lie on top of other entities. If you do get an error, observe your sketch to see if there are any obvious problems. If necessary, try to delete entities created with the Use Edge option. You can use the Undo and Redo options to help with the manipulation. A final work-around solution would be to sketch the section's inside arc using the Arc option instead of the Use Edge option.

STEP 10: **Enter a BLIND depth of .500.**

STEP 11: **Preview the protrusion, then select OKAY on the dialog box.**

STEP 12: **If necessary, select DONE on the Feature menu.**

STEP 13: **Select the SET UP >> NAME command from the Menu Manager.**

STEP 14: **On the Model Tree, select the Protrusion feature that defines the previously created extruded feature and rename it *EXTRUDE1*.**

This step renames the last feature. Use the Name option to rename the feature *EXTRUDE1* (Figure 5–22).

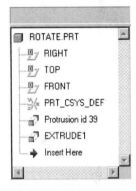

Figure 5-22   Naming a feature

CREATING A COAXIAL HOLE

This segment of the tutorial will create the Coaxial Hole shown in Figure 5–23. To create the hole, the datum axis shown will have to be created first. This axis will be created with the Datum Axis command and the Through Cylinder constraint option.

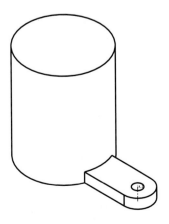

Figure 5-23   Coaxial Hole feature

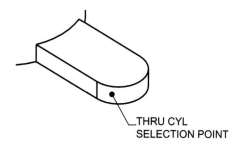

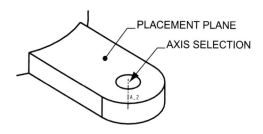

Figure 5-24   Cylinder selection

Figure 5-25   Coaxial Hole creation

STEP 1:   Select the DATUM AXIS icon on the Datum toolbar.

STEP 2:   Select THRU CYL as the constraint option.

The Thru Cylinder constraint option will place a datum axis through the center of an existing cylindrical feature. The cylindrical feature can be any solid feature with a rounded surface (e.g., shaft, round, extruded arc, etc.).

STEP 3:   Select the *Extrude1* feature in the location shown in Figure 5–24.

STEP 4:   Create the Coaxial Hole shown in Figure 5–25.

Use the following parameters when creating the hole:

- Create a Straight hole.
- Use a Hole diameter of 0.500.
- Create the Hole with a Thru All depth one parameter.
- Select the previously created datum axis as the primary reference.
- Use the Coaxial placement option.

MIRRORING THE EXTRUDED FEATURE

This segment of the tutorial will mirror the extruded Protrusion, Coaxial Hole, and Datum Axis features (Figure 5–26). The Copy >> Mirror option requires the selection of a mirror plane. In this tutorial, one of Pro/ENGINEER's default datum planes will be used as this plane.

STEP 1:   Select FEATURE >> COPY.

STEP 2:   Select the MIRROR option from the Copy Feature menu.

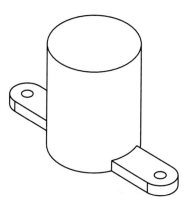

Figure 5-26   Mirrored features

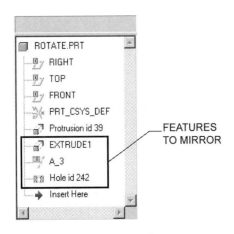

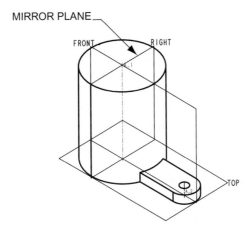

Figure 5-27   Feature selection   **Figure 5-28**   Mirror plane selection   **Figure 5-29**   Model tree

Step 3:   **Select DEPENDENT on the Copy Feature menu, then select DONE.**

With the Dependent option, the copied features' dimensions are dependent upon its parent features' dimensions. If a dimension on a parent feature changes, the corresponding dimension on its child will change.

Step 4:   **On the Model Tree, select the EXTRUDE1 feature, the Datum Axis feature, and the Coaxial Hole feature (Figure 5–27).**

The last three features on the Model Tree will be copy-mirrored. Optionally, you can select the features on the work screen.

Step 5:   **Select DONE SEL on the Get Select menu.**

Select Done Select to exit the feature selection process.

Step 6:   **Select DONE on the Select Feature menu.**

The Select Feature menu provides an option for selecting additional features to be copied. Select Done to end the selection process.

Step 7:   **As shown in Figure 5–28, select datum plane RIGHT as the plane to mirror the selected features about.**

After selecting the plane to mirror about, the mirrored grouped features will be created.

Step 8:   **Observe your Model Tree.**

Notice the group feature added to the Model Tree (Figure 5–29). By selecting the + icon next to the feature on the Model Tree, the group will be expanded to reveal the elements of the group.

ROTATING THE EXTRUDED FEATURE

The Copy command has an option for rotating features. Unlike the Pattern command, the Rotate option can array multiple features simultaneously. This segment of the tutorial will rotate the Protrusion, Hole, and Axis features 90 degrees to create the grouped feature shown in Figure 5–30.

Step 1:   **Select FEATURE >> COPY.**

Step 2:   **Select the MOVE option from the Copy Feature menu.**

Other options available under the Copy Feature menu include New Refs, Same Refs, and Mirror.

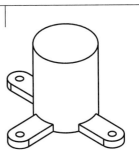

Figure 5-30   Rotated features

STEP 3: **Select INDEPENDENT >> DONE.**

The Independent option will make the copied features' dimension values independent from its parent features' values. If a parent feature's dimension value changes, the copied feature's corresponding dimension value will not change.

STEP 4: **On the Model Tree, select the EXTRUDE1 feature, the original Coaxial Hole, and the Axis locating the Coaxial Hole (Figure 5–31).**

Select the features that were mirrored in the previous section of this tutorial.

STEP 5: **Select DONE SEL on the Get Select menu.**

STEP 6: **Select DONE on the Select Feature menu.**

STEP 7: **Select ROTATE on the Move Feature menu.**

The Rotate option will copy a feature about an edge, axis, or curve.

STEP 8: **Select CRV/EDG/AXIS, then select the Axis defining the center of the base revolved protrusion (Figure 5–32).**

As shown in Figure 5–32, the axis used to define the center of the revolved protrusion feature will be used as the center of the rotation. On the work screen, select this axis.

STEP 9: **Select OKAY to accept the Direction for Translation (Figure 5–32).**

On the work screen, an arrow points in the direction of translation. Pro/ENGINEER uses the right-hand rule to determine this direction. Using your right-hand, with the thumb pointed in the direction of the arrow, the copy will rotate the direction that the fingers of your hand are pointing.

STEP 10: **Enter 90 as the Rotation Angle.**

STEP 11: **Select DONE MOVE on the Move Feature menu.**

STEP 12: **Select DONE on the GP VAR DIMS menu.**

The dimensions from the features that are being copied can be varied during the copy process. The GP VAR DIMS menu provides the option of selecting dimensions to vary during the copy.

STEP 13: **Select OKAY on the dialog box.**

STEP 14: **If necessary, select DONE to exit the feature menu; then save the part.**

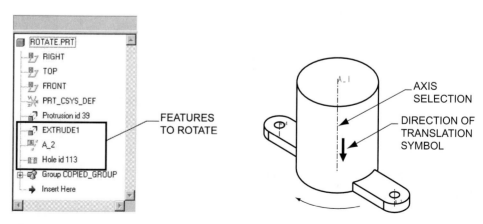

Figure 5–31   Feature selection   **Figure 5–32**   Axis selection

ADDING RELATIONS

This segment of the tutorial will create a dimensional relationship that will make the distance from the center of the base protrusion to the center of the first axial hole equal to the height of the base protrusion. In Figure 5–33, dimension d12 will be set equal to dimension d1.

STEP 1: **Select the RELATIONS option from the Menu Manager.**

STEP 2: **On the work screen, select the base revolved Protrusion and the EXTRUDE1 feature.**

The dimensions shown in Figure 5–33 should appear similar to your model. Within the Relations menu, dimensions are shown symbolically. The symbols displayed in the illustration may or may not match your symbols. When following this tutorial, use the corresponding symbols from your model.

From Figure 5–33, you will add a relation that will make the dimension displayed with symbol d12 equal to the dimension displayed with the symbol d1.

STEP 3: **Select the ADD option on the Relations menu.**

STEP 4: **With the symbols shown in Figure 5–33, add an equation that will make the d12 dimension equal to the d1 dimension.**

For the symbols shown in Figure 5–33, add the equation:

$$d12 = d1$$

STEP 5: **Select ENTER on the keyboard to quit the Add option.**

STEP 6: **Select DONE on the Model Relations menu.**

STEP 7: **Select REGENERATE on the Part menu.**

After regenerating, notice how the EXTRUDE1 feature and the mirrored feature lengthen (Figure 5–34). Why did the second copy-rotated feature not lengthen with the other two features? The Independent option used in the creation of the rotated copy made its dimensions independent of its parent feature's dimensions.

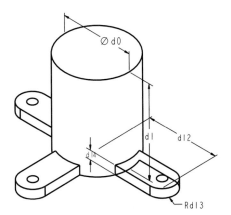

Figure 5-33   Dimension symbols

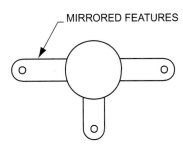

Figure 5-34   Finished features

MANIPULATING TUTORIAL 2

This tutorial will create the part shown in Figure 5–35. The primary objective of this tutorial is to demonstrate the Group command and the ability to pattern grouped features. Covered in this tutorial will be the following topics:

- Creating a Through >> Axis datum plane.
- Creating an extruded boss feature.
- Creating a coaxial hole and fillet.
- Grouping features.
- Patterning a group.
- Creating a conditional relation.

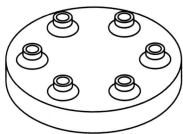

Figure 5-35   Finished part

CREATING THE BASE PROTRUSION

The base feature for this tutorial is shown in Figure 5–36. Using Pro/ENGINEER's default datum planes and the following requirements, create this feature.

- Revolve the section shown in Figure 5–36.
- Use Pro/ENGINEER's default datum planes.
- Sketch the revolved protrusion on datum plane FRONT.
- 360-degree angle of revolution.

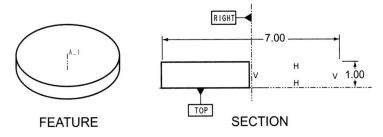

Figure 5-36   Base protrusion

CREATING A THROUGH >> AXIS DATUM PLANE

This tutorial will create a datum plane through the center axis of the base feature and at an angle to datum plane FRONT. The boss features shown in Figure 5–35 are patterned groups. To create a revolved patterned group, an angular dimension used within the definition of a feature in the group must be used as the leader dimension. This segment of the tutorial will create the angular datum plane shown in Figure 5–37. The angle defining this datum plane will be the leader dimension for the pattern.

STEP 1: Select VIEW >> DEFAULT from the Menu Bar.

STEP 2:   Select the DATUM PLANE icon.

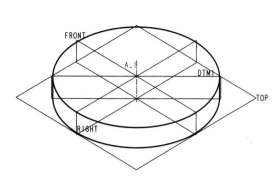

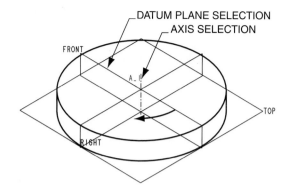

Figure 5–37   Through >> Axis datum plane

Figure 5–38   Datum creation

STEP 3: Select the THROUGH option from the Datum Plane menu, then select the axis at the center of the base protrusion (Figure 5–38).

The Through option will construct a datum plane through an edge, axis, curve, point, vertex, plane, or cylinder.

STEP 4: Select the ANGLE option from the Datum Plane menu, then select datum plane FRONT (Figure 5–38).

When using the Through option, with the exception of the Through >> Plane constraint, a paired constraint option is required. The Angle constraint option will construct the datum plane at an angle to a selected plane. In combination, the Through >> Axis and Angle constraint options will construct this new datum plane through the center axis and at a specified angle to datum plane FRONT.

STEP 5: Select DONE from the Datum Plane menu.

STEP 6: Select ENTER VALUE from the Offset menu.

STEP 7: In Pro/ENGINEER's textbox, enter −45 as the angular value.

On the work screen, Pro/ENGINEER will graphically display the direction of rotation for the angular datum plane. In this example, entering a value of −45 will create the new datum plane at a 45-degree angle to FRONT.

CREATING THE BOSS FEATURE

This segment of the tutorial will create the Extruded Protrusion shown in Figure 5–39. This protrusion will serve as the first feature in the group to be patterned.

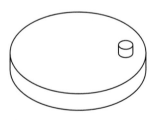

STEP 1: Select FEATURE >> CREATE >> PROTRUSION.

STEP 2: Select EXTRUDE >> SOLID >> DONE >> ONE SIDE >> DONE.

STEP 3: Select the top of the part as the sketching plane, then accept the default extrusion direction.

Figure 5–39   Boss feature

STEP 4: Select TOP, then pick DTM1 to orient this plane toward the top of the sketcher environment (Figure 5–40).

INSTRUCTIONAL POINT   The orientation of datum plane DTM1 toward the top of the sketcher environment is an important step of this tutorial. Without this proper orientation, you will get an error message when patterning the group later in this tutorial. As shown in Figure 5–40, datum plane DTM1 should be horizontal on the work screen. If datum plane DTM1 is not horizontal, restart this segment of the tutorial.

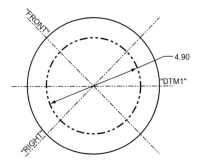

Figure 5-40   Construction circle

STEP 5: Use the References dialog box to specify datum planes RIGHT, FRONT, and DTM1 as references.

STEP 6: 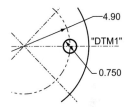 Use the CIRCLE icon to create the circle shown in Figure 5–40.

STEP 7: With the previously created circle still picked, on Pro/ENGINEER's Menu Bar, select EDIT >> TOGGLE CONSTRUCTION.

Elements must be picked before you can turn them into construction entities. If the circle is not selected, use the Pick icon to select it. Elements created as Construction entities will not extrude with geometry entities. This construction circle will be used to align and locate the extruded feature.

STEP 8: Modify the construction circle's diameter to equal a value of 4.90.

STEP 9: Use the CIRCLE option to create the circle entity shown in Figure 5–41.

Align the center of the new circle at the intersection of the construction circle and datum plane DTM1. Modify the Circle's diameter to equal a value of 0.750.

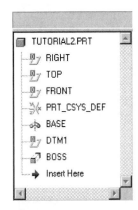

Figure 5-41   Circle creation

STEP 10: Exit the sketcher environment and extrude the Protrusion a BLIND distance of .500.

STEP 11: Preview the Protrusion, then select OKAY on the Feature Definition dialog box.

STEP 12: Select DONE to exit the Feature menu, then select the SET UP menu option.

The next three steps will give each Protrusion feature a more descriptive name.

STEP 13: Select the NAME option from the Part Setup menu.

STEP 14: On the Model Tree, select the first Protrusion feature and rename it *BASE* (Figure 5–42).

This will allow for the renaming of this feature to make it more descriptive on the Model Tree.

STEP 15: Use the NAME command to rename the last Protrusion feature *BOSS* (Figure 5–42).

STEP 16: Save your part.

Figure 5-42   Model tree

CREATING A COAXIAL HOLE AND ROUND

This segment of the tutorial will create the Hole and Round features shown in Figure 5–43. These two features will be combined with the Boss feature and datum plane DTM1 to form a group.

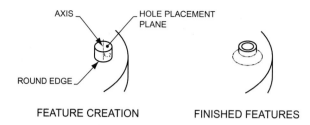

Figure 5-43   Hole and Round features

STEP 1: **Create a COAXIAL HOLE as shown in Figure 5–43 by using the following options:**

- Use the Hole command.
- Create a Straight hole.
- The hole will have a diameter value of 0.500.
- Create the hole using the Thru All depth option.
- Use the BOSS feature's axis as the primary reference.
- Place the hole with the Coaxial option.

STEP 2: **Create a ROUND as shown in Figure 5–43 using the following options:**

- Use the Round command.
- Create a Simple Round.
- Use a Constant radius value of 0.250.
- Use the Edge Chain and Tangent Chain options to select the base of the Boss feature.

GROUPING FEATURES

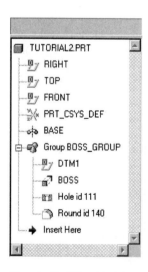

Figure 5-44   Model tree

This section of the tutorial will group datum plane DTM1 with the Boss, Hole, and Round features. When features are grouped, they are turned essentially into one feature. The design of this part requires these features to be arrayed about a bolt-circle. Since the normal Pattern command arrays one feature at a time, these features have to be grouped first (Figure 5–44).

STEP 1: **Select FEATURE >> GROUP.**

STEP 2: **Select LOCAL GROUP on the Create Group menu.**

Local groups are available only within the model in which they were created.

STEP 3: **Enter *BOSS_GROUP* as the Name of the Local Group.**

STEP 4: **On the Model Tree (Figure 5–45), select datum plane DTM1, the BOSS feature, the Hole feature, and the Round feature.**

Each feature could be selected on the work screen. Since features must be adjacent to each other in the order of regeneration, it is often easier to pick features on the Model Tree.

STEP 5: **Select DONE on the Select Feature menu, then observe the changes to the Model Tree.**

Your Model Tree should appear as shown in Figure 5–44.

STEP 6: **Select DONE/RETURN on the Group menu.**

STEP 7: **Save your model.**

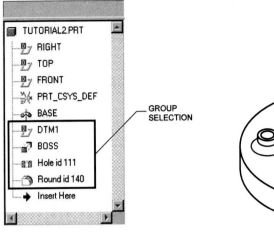

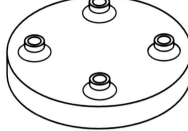

Figure 5-45   Feature selection

Figure 5-46   Patterned group

PATTERNING THE BOSS GROUP

The normal Pattern command can only pattern one feature at a time, and does not work with grouped features. The Group menu provides an option for patterning groups. This option will be used to create the rotational pattern shown in Figure 5–46.

STEP 1: Select FEATURE >> GROUP.

STEP 2: Select PATTERN on the Group menu.

The Group menu provides an option to pattern grouped features. The group Pattern option works similarly to the normal Pattern option. Unlike normal patterns, the Group >> Pattern option only allows Identical patterns. Due to this, patterned group instances must lie on the same placement plane, instances cannot intersect, and instances cannot intersect an edge.

STEP 3: On the Model Tree or on the work screen, select the *BOSS_GROUP* feature.

STEP 4: For the Leader dimension in the first direction, pick the 45-degree dimension used to define the angle of datum plane DTM1 (Figure 5–47).

Select the 45-degree dimension shown in Figure 5–47 as the first direction leader dimension. This dimension is the angular reference dimension obtained from the creation of datum plane DTM1. Since DTM1 was included in the group, this angular dimension is available for varying under the Group >> Pattern option. Since rotational patterns require an angular dimension, this is the necessary dimension required in the first direction of the pattern.

STEP 5: Enter 90 as the leader dimension's increment value in the first direction.

Entering a 90 as the increment value will create each instance of the pattern 90 degrees apart.

STEP 6: Select DONE from the Exit menu.

Selecting Done will stop the selection of dimensions in the first direction.

STEP 7: Enter 4 as the number of instances in the first direction.

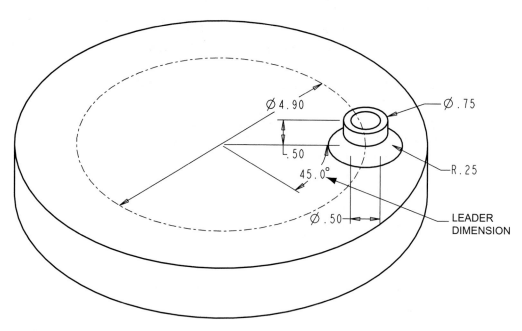

Figure 5-47   Leader dimension selection

STEP 8: **Select DONE to create the pattern.**

This group will be patterned in one direction only. Due to this, the selection of a second leader dimension is not necessary. Selecting Done from the exit menu will create a pattern of the group.

STEP 9: **Select DONE/RETURN from the Group menu.**

STEP 10: **Save your part.**

ESTABLISHING A CONDITIONAL RELATIONSHIP

One of the powerful capabilities of a Feature-Based/Parametric modeling package is its ability to incorporate design intent into a model. One of the ways that intent can be built into a model is through the creation of dimensional relationships. A relation is an explicit mathematical relationship that exists between two dimensions. The flexibility of the Relations command allows conditional statements to be built into a relationship's equation. This tutorial will create a dimensional relationship that utilizes a conditional statement.

The current diameter of the base feature in this tutorial is 7 inches. This portion of the tutorial will add a conditional relationship that will drive the number of BOSS_GROUP features within the rotational pattern. The following design intent applies.

- A base feature diameter of 5 inches or less will have 4 equally spaced boss features.
- A base feature diameter over 5 inches, but less than or equal to 10 inches, will have 6 equally spaced boss features.
- A base feature diameter over 10 inches will have 8 equally spaced boss features.
- The centerline diameter of the patterned groups will be 70 percent of the diameter of the base feature's diameter.

As shown in Figure 5–48, the following dimensions with matching symbols will be used. Your symbols may be different.

- **Base feature diameter (d0)** This is the diameter value of the base feature. Initially, it is set to a value of 7 inches.

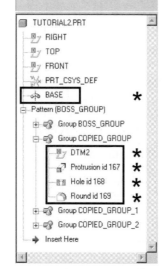

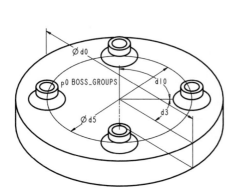

Figure 5-48   Dimension symbols   **Figure 5-49**   Model tree

- **Number of grouped boss features (p0)**   This is the number of instances that the grouped boss was patterned. Initially, there are four instances.
- **Diameter of the pattern (d5)**   This is the diameter value that defines the location of the first Boss feature. Initially, this value is set to 4.90 inches.
- **Increment value of the pattern (d10)**   This is the number of degrees incremented between each instance of the pattern. Initially, this value is set to 90 degrees.

STEP   1:   Select the RELATIONS option from the Part menu.

STEP   2:   On the Model Tree, expand the Pattern feature and the second Group feature by selecting the + sign to the left of the required feature's name (Figure 5–49).

Selecting the + sign next to a feature's name will reveal components, features, and elements defining the feature.

STEP   3:   On the Model Tree, select the BASE Protrusion feature and each feature defining the second BOSS_GROUP group (Figure 5–49).

Selecting features of the first grouped feature will not reveal the dimension symbol defining the pattern increment value. Before adding relations, make sure that the dimensions shown in Figure 5–48 are revealed on your work screen.

STEP   4:   Select the ADD option on the Relations menu.

STEP   5:   Enter the relation statements shown below.

Enter each statement line by line as shown. Selecting the Enter key at the end of each statement will allow for the entering of a new line. An alternative is to use the Edit Relations (Edit Rel) option (Figure 5–50). Make sure that you use the dimension symbols associated with your model. Your symbols may be different from those shown in Figure 5–48 and from those shown in table 5–2.

INSTRUCTIONAL POINT   While entering relation equations in the textbox, you may get an error resulting in an apparent frozen textbox. In this case, enter the statement *ENDIF* into the textbox. This error is a result of an unsatisfactory ending to a conditional statement.

Table 5-2 Statement line table

Statement Line	Meaning
d5 = d0*.70	Diameter of the bolt-circle is 70% the diameter of the flange
IF d0 <= 5	If the diameter of the flange is less than or equal to 5.00
p0 = 4	The number of holes is equal to 4
d10 = 90	The angular increment value is equal to 90 degrees
ENDIF	End of the definition
IF (d0 > 5) & (d0 <= 10)	If the flange diameter is greater than 5.00 and less than or equal to 10.00
p0 = 6	The number of holes is equal to 6
d10 = 60	The angular increment value is equal to 60 degrees
ENDIF	End of the definition
IF d0>10	If the flange diameter is greater than 10.00
p0 = 8	The number of holes is equal to 8
d10 = 45	The angular increment value is equal to 45 degrees
ENDIF	End of the definition

Figure 5-50 Edit Relations option

STEP 6: Select the REGENERATE option on the Part menu

For relational changes to take effect, you have to regenerate the model.

STEP 7: Use the MODIFY >> VALUE option to change the base diameter dimension to a value of 12.00

How many boss groups should the part have after regeneration?

INSTRUCTIONAL NOTE For this particular part, changing the flange diameter to a value of 4.00 or less will produce a regeneration error. The Group >> Pattern option defaults to an Identical pattern option. Instances of an Identical pattern cannot intersect each other or the placement plane's edge. In this case, making the flange too small can violate both requirements.

STEP 8: Regenerate the model.

STEP 9: MODIFY the base diameter dimension to a value of 9.00, then REGENERATE the model.

You should have six boss groups.

STEP 10: MODIFY the base diameter dimension to a value of 4.75, then REGENERATE the model.

STEP 11: Save your part.

PROBLEMS

1. Using Pro/ENGINEER's Part mode, model the part shown in Figure 5–51. Construct the part using the following order of operations:

    a. Create the base geometric feature as a revolved Protrusion. Include the hole in the section.

    b. Model one leg feature (including the hole).

    c. Use the Copy >> Move >> Rotate option to create three instances of the leg feature.

    d. Create the remaining three leg instances.

2. Model the part shown Figure 5–52.

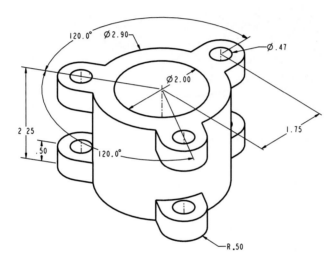

Figure 5-51 Problem one

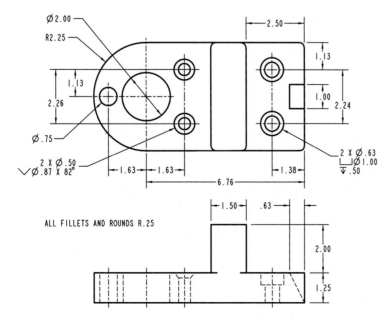

Figure 5-52 Problem two

3.   Model the part shown Figure 5–53. Use the Copy-Rotate option to create the multiple instances of the Cut and Hole features.

4.   Model the part shown in Figure 5–54.

5.   Model the part shown in the drawing in Figure 5–55.

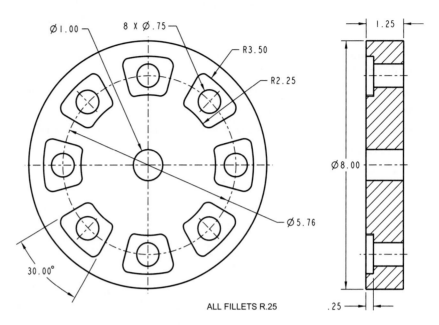

Figure 5-53   Problem three

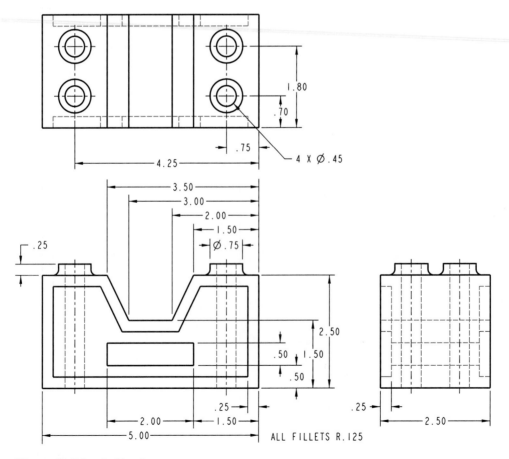

Figure 5-54   Problem four

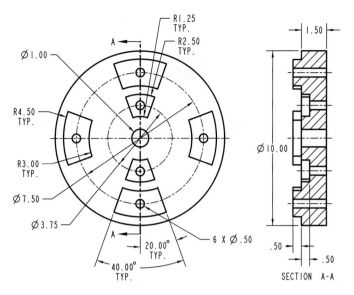

Figure 5-55   Problem five

QUESTIONS AND DISCUSSION

1. Describe the difference between the Pattern command and the Copy command. Describe some situations when Copy would be used in place of Pattern.

2. Describe the difference between the Copy >> Same Refs option and the Copy >> Move >> Translate option. Can you think of a situation when you might use one option over the other?

3. Describe the difference between copying features independently and copying features dependently.

4. Write relation statements for the following situations:

   a. Dimension d0 is equal to two times dimension d1.

   b. Dimension d0 is equal to the sum of dimension d1 and d2 divided by d3.

   c. Dimension d0 is equal to the square root of d1.

6

CREATING A PRO/ENGINEER DRAWING

Introduction

Drawing mode is used to produce detailed engineer drawings of parts and assemblies. Drawing mode has the capability to produce orthographic views to include section views and auxiliary views. This chapter introduces the fundamentals behind Drawing mode and how to produce an annotated multiview drawing. Upon finishing this chapter, you will be able to

- Within Drawing model, create orthographic views of an existing model.
- Manipulate the settings of a drawing by changing Drawing Setup file options.
- Create a detailed view of a model.
- Apply parametric and nonparametric dimensions to a drawing.
- Show and erase entities such as Geometric Tolerances, Centerlines, and Datums.

DEFINITIONS

Associative dimension   A parametric dimension available for viewing and/or modification in multiple modules of Pro/ENGINEER.

Drawing setup file   A text file used to establish many of Drawing mode's default settings. As an example, the default text height for a drawing is set in the drawing setup file.

Format   A Pro/ENGINEER module used to create drawing borders and title blocks. Formats can be added to a Pro/ENGINEER drawing.

Multiview drawing   The use of multiple orthographic views to graphically display and communicate an engineering design.

Nonparametric dimension   A dimension created in Drawing mode that is not used to construct a part or assembly. Nonparametric dimension values cannot be modified.

Parametric dimension   A dimension used to define a part or assembly feature. Parametric dimensions can be modified or redefined.

DRAWING FUNDAMENTALS

Pro/ENGINEER is a fully integrated and associative engineer design package. Since it is an integrated package, an array of design, engineering, and manufacturing tools is available. Components of a design can be modeled within Part mode, grouped in Assembly

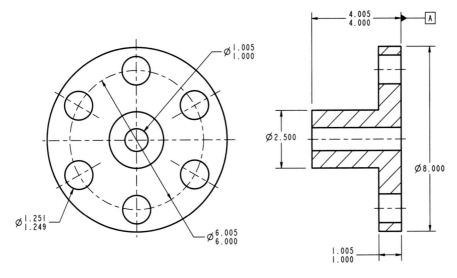

Figure 6-1   Pro/ENGINEER drawing

mode, simulated and tested in Pro/MECHANICA, with machining code developed in Manufacturing mode. The two-directional associativity that exists between modes allows changes made in one mode to be reflected in another. When Pro/ENGINEER is combined with a strong computer network system, a true "paperless manufacturing" environment can be established.

Two-dimensional orthographic drawings were once considered one of the initial steps in the design process. A traditional approach to engineering design might require engineering drawings to be produced first, followed by engineering analysis and CNC code production. With Pro/ENGINEER, engineering drawings are considered "downstream" applications that occur after part modeling and analysis.

Pro/ENGINEER's basic package provides a module for creating orthographic drawings (Figure 6–1). Within Drawing mode, detailed multiview drawings can be created from existing models. Dimensions used to create a part (referred to as **parametric dimensions**) can be revealed in Drawing mode to document a design. Parametric dimensions can be modified in Drawing mode with the changes reflected in other modules of Pro/ENGINEER. Options are available for creating section views, detailed views, and auxiliary views. There are options for creating notes, leaders, **nonparametric dimensions,** and tables. Additionally, within Drawing mode, there exist a variety of two-dimension drafting tools.

Drawing mode can be used to create detailed drawings from existing parts and assemblies. When creating a new drawing, Pro/ENGINEER provides an option for selecting the model from which to create the drawing. Also, additional models can be added to a drawing from within Drawing mode. Multiple views of a model can be added to a drawing and annotations applied. Pro/ENGINEER provides existing sheet formats (e.g., A, B, C, D, etc.) with detailed title blocks and borders. Format mode can be used to create additional sheet formats.

DRAWING SETUP FILE

Drawing mode uses **Drawing Setup Files** (DTL) to control the appearance of drawings. Pro/ENGINEER comes with default settings for a variety of drawing parameter options. Examples of parameters include text height, arrowhead size, arrowhead style, tolerance display, and drawing units. Default values can be changed permanently or for individual drawings. Multiple drawing setup files can be created and stored for later use. Pro/ENGINEER's

Table 6-1   Common drawing setup file options

Option/Description	Default Value (Optional Value)
crossec_arrow_length Controls the length of cutting-plane line arrowheads.	0.1875
crossec_arrow_width Controls the width of cutting-plane line arrowheads.	.06250
dim_leader_length Controls the length of a dimension line when the dimension line arrowheads fall outside of the extension lines.	0.5000
draw_arrow_length Sets the length of dimension arrowheads.	0.1875
draw_arrow_style Sets the arrowhead style.	closed (open or filled)
draw_arrow_width Sets the width of dimension arrowheads.	0.0625
drawing_text_height Sets the height of text in a drawing.	0.15625
drawing_units Sets units for a drawing.	inch (foot, mm, cm, or m)
gtol_datums Sets the display of geometric tolerance datum symbols.	Std_ansi (std_iso, std_jis, or std_ansi_mm)
leader_elbow_length Sets the length of a leader's elbow.	0.2500
radial_pattern_axis_circle Controls the display of rotational pattern features. Set to *yes* produces a bolt-circle centerline.	no (yes)
text_orientation Sets the orientation of dimension text.	horizontal (parallel or parallel_diam_horiz)
text_width_factor Sets the width factor for text.	0.8000 (0.25 through 8)
tol_display Sets the display of tolerance values.	no (yes)

default drawing setup file (*prodetail.dtl*) can be found in the *<pro_engineer load point>\text*directory.

ADVANCED >> DRAW SETUP

New DTL Files are created or current drawing settings are changed with the Advanced >> Draw Setup option. Pro/ENGINEER utilizes an Options dialog box similar to the configuration file's Options dialog box for drawing setup changes (see Chapter 1). The configuration file option *drawing_setup_file* can be used to establish a specific DTL file. If this option is not set, Pro/ENGINEER uses the default DTL file. The configuration file option *pro_dtl_setup_dir* can be used to set the directory that Pro/ENGINEER searches for DTL files. Table 6–1 provides a list of common DTL file options with default values.

SHEET FORMATS

A Format is an overlay for a Pro/ENGINEER drawing. It can include a border, title block, notes, tables, and graphics. A sheet format has the file extension **.frm*. Pro/ENGINEER provides a variety of predefined standard formats for use with ANSI and ISO sheet sizes (e.g., A, B, C, and D size sheets). These standard formats can be modified to produce a customized format. Additionally, Pro/ENGINEER's Format mode can be used to create a new sheet format. Figure 6–2 shows an example of an A size sheet format. A library of standard

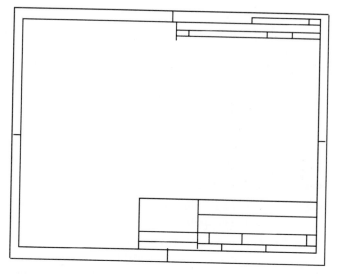

Figure 6-2   ANSI A size format and title block

sheets can be created. The configuration file option *pro_format_dir* can be used to specify the directory path where standard sheets are stored.

CREATING A NEW DRAWING

The Drawing mode option from the New dialog box is used to create new drawings. When Drawing mode is selected and a file name entered, Pro/ENGINEER introduces the New Drawing dialog box (Figure 6–3). If a part or assembly is currently active in session memory, Pro/ENGINEER defaults to this part or assembly as the model from which to create the drawing. An option is available for browsing to find other existing models. The New Drawing dialog box provides the option for specifying a standard sheet size or for retrieving an existing format.

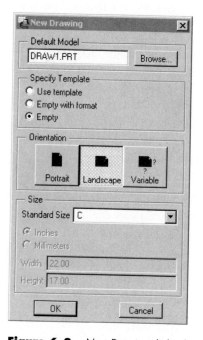

Figure 6-3   New Drawing dialog box

Three types of items can be added to a drawing: Formats, 2D draft entities, and Model views. Formats are placed in a drawing using the Sheets >> Format >> Add/Replace option from the Drawing menu. Pro/ENGINEER provides an Open dialog box for browsing to find an appropriate format. Multiple sketching tools allow for the creation of 2D draft entities. Other options are available for adding dimensions and notes. Model views can be added to a drawing with the Views >> Add option. General, section, projected, auxiliary, revolved, and detail views can be added.

The following instructions show step-by-step how to create a new drawing. Several options are available that can vary the method for creating a new drawing. This process will allow for the creation of a new drawing without specifying a standard sheet size.

STEP 1: Select FILE >> NEW.

STEP 2: Select Drawing mode from the New dialog box, enter a name for the new drawing, then select OKAY. (Use the default template file)

STEP 3: On the New Drawing dialog box, select the DEFAULT MODEL from which to create the drawing.

Pro/ENGINEER will default to the current active model. You can use the Browse option to search for any existing object.

STEP 4: Select the EMPTY button from the Specify Sheet option.

The Empty option will require the defining of a sheet size and orientation. Alternately, the Use Template option is usable for the selection of an existing format, while the Empty-with-Format option is used to create a drawing without an established size or sheet format.

STEP 5: Select a sheet Orientation and a sheet Size, then select OKAY.

The Empty option requires the selection of a sheet orientation and size. A Portrait, Landscape, or Variable orientation is available. Standard sheet sizes and user-defined sheets sizes are also available.

DRAWING VIEWS

Pro/ENGINEER's Drawing mode provides options for creating a variety of orthographic views to include section views, auxiliary views, detailed views, revolved views, broken views, and partial views. As many views as necessary to fully describe a model can be added to a drawing sheet. Once a view is inserted into a drawing, parameters associated with the model, to include dimension values, can be shown. Views are also fully associated. This allows any model changes made in a drawing to be reflected in the part or assembly model. Additionally, changes in one view of a drawing will reflect accordingly in all views of the model.

THE VIEWS MENU

The Views menu option is used to create views of an existing Pro/ENGINEER model. The Views menu has options for manipulating and modifying existing views. The following menu options are available.

ADD VIEW

The Add View option is used to create new views. The first view created must be a General view. From a General view, other views of the model can be added.

MOVE VIEW

The Move View option is used to move views on the work screen. When placing a view, Pro/ENGINEER requests a Center Point for the drawing view. The view is set initially at this location. The Move View option can be used to reposition this or any other view.

MODIFY VIEW

The View Modify menu option provides a variety of tools for modifying a view. The following is a partial list of the available options.

- **View Type** The View Type option is available to change the type of view. As an example, a Projection view can be changed to a General or Auxiliary view.

- **Change Scale** This option is used to change the scale of a nonchild view. The view cannot be a child view of another view, and the view must have been inserted with a specific, user-defined scale value.

- **View Name** Pro/ENGINEER provides each view with a unique name. This option is used to rename a view.

- **Reorient** When inserting a view, Pro/ENGINEER provides the option of orienting the view. As an example, a view can be oriented to allow for a proper front view vantage point. The Reorient option allows this orientation to be changed.

- **X-Section** This option allows a cross section to be replaced.

- **Z-Clipping** The Z-Clipping option allows for the exclusion of all graphics on a view behind a selected plane. This option is advantageous for views with background graphics that can clutter the drawing.

ERASE VIEW

By erasing a view, you can temporarily remove it from the drawing screen. Erasing a view removes it from the regeneration but does not affect any other view, including child views. To return a view to the drawing screen, use the Resume View option.

DELETE VIEW

The Delete view option permanently removes a view from the drawing. Views that are parents of other views cannot be deleted.

RELATE VIEW

The Relate View option is used to assign draft entities to a selected view. As an example, a note might be placed into a drawing using the Create >> Note option. This note can be assigned to a specific view with the Relate View option.

DISPLAY MODE

This menu option is used to modify the display of a selected view. The following options are available:

- **View Display** This option is used to change the display of lines on a drawing view. A view's hidden lines can be specified as Wireframe, Hidden, or No Hidden. Additionally, tangent edge lines may be specified to display as a solid line, a centerline, a phantom line, a dimmed line, or set to not to be displayed.

- **Edge Display** The Edge Display option is similar to the View Display option. With the Edge Display option, individual hidden or tangent edges can be changed.

- **Member Display** This option is used to control the display of assembly views.

DRAWING MODELS

Multiple models can be accessed within one drawing. The Drawing Models (Dwg Models) option allows additional models to be added to the current drawing. The Set option allows a specific model to be set as the active model in the drawing.

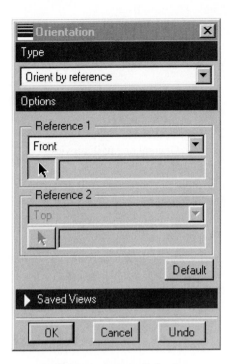

Figure 6-4   Orientation dialog box

VIEW TYPES

Pro/ENGINEER provides a variety of view types to serve the documentation needs of a model. The following is a list of the available types.

GENERAL VIEWS

The General view is the basic view type available. It is required as the first view placed into a drawing and is used by other types as a parent view. General views require a user-defined orientation. A General view is first placed into a drawing using Pro/ENGINEER's current default view orientation. The Orientation dialog box (Figure 6–4) is used to orient the view to match orthographic view projection requirements. In Figure 6–5, the Front view of the object was inserted as a General view.

PROJECTION VIEWS

The Projection view is an orthographic projection from a General view or from an existing view. Projection views follow normal lines of projection according to conventional drafting standards. As an example, a Projection view would be used to create a right-side view off of an existing front view. Projection views become child views of the view from which they are projected. In Figure 6–5, the Top and Right-Side views were created as Projection views.

AUXILIARY VIEWS

Auxiliary views are used to project a view when normal lines of projection will not work. They are used to show the true size of a surface that cannot be shown from one of the six primary views. Auxiliary views are projected from a selected edge or axis. The auxiliary view shown in Figure 6–5 is projected off the Front view.

DETAILED VIEWS

Often, features are too small to describe fully with a standard projection view. In such a case, it is common practice to enlarge portions of a drawing to allow for more accurate detailing. Figure 6–5 shows an example of a detailed view.

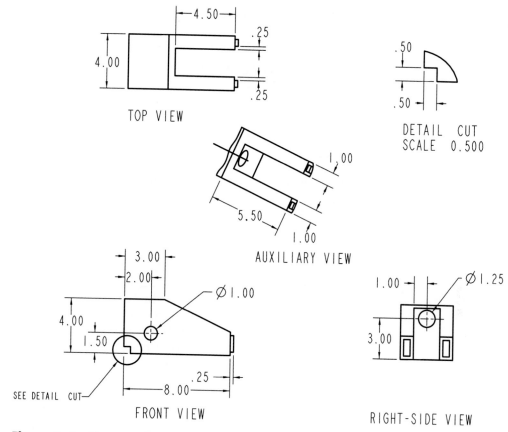

Figure 6-5   View examples

REVOLVED VIEWS

Revolved views are used to show the Cross Section of a part or feature. A Cross Section is required and can be retrieved or constructed from within the Revolve option. Once selected, the Cross Section is revolved 90 degrees from the cutting plane. A Revolved view can be either a Full view or a Partial view.

VIEW VISIBILITIES

In addition to view types such as General, Projection, and Auxiliary, Drawing mode provides the option of controlling the visibility of selected portions of a view. The following View Visibility options are available.

HALF VIEWS

Often, it is not necessary to show an entire model in a view. A good example would be a symmetrical object. Half views remove the portion of a model on one side of a selected datum plane or planar surface. Figure 6–6 shows an example of a Half view. Half views can be used with General, Projection, and Auxiliary views only.

PARTIAL VIEWS

It is not necessary to document an entire view when only a small portion of a feature needs to be detailed. The Partial View option allows a selected small portion of a view to be created. The area to be revealed is enclosed in a sketched spline. Figure 6–6 shows an example of a Partial Section view. Unlike Detailed views, Partial views must follow normal lines of projection. Partial views can be used with General, Projection and Auxiliary views only.

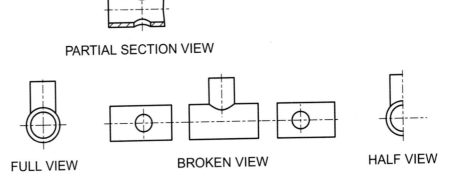

Figure 6-6   View types

BROKEN VIEWS

As with Partial and Half views, in many views it is not necessary to show the entire object. As shown in Figure 6–6, Broken views are used often with long, consistent cross sections. With a Broken view, multiple horizontal or vertical breaks may be used if necessary. The Drawing Setup file option *broken_view_offset* is used to set the offset distance between portions of a view. The Move View option can be used to move a portion of a Broken view.

SINGLE-SURFACE VIEWS

A single-surface view is a projection of an individual part surface. They may be used with any view type except Detailed and Revolved. When creating a single-surface view, only the selected surface is revealed. The Of Surface option is used to create a single-surface view.

SETTING A DISPLAY MODE

Views may be displayed as Wireframe, Hidden, or No Hidden. Views set with the Hidden option will print dashed hidden lines. By default, the display mode of a view is dependent upon the display mode set for the object. An object's display mode can be set on the tool-bar or in the Environment dialog box. Individual display modes can be set for each view of a drawing using the Views >> Disp Mode option. The display of edges and assembly members can be manipulated with the Display Mode option also.

Tangent edges can be displayed in a variety of styles on a model. The Environment dialog box setting *Tangent Edges* is used to set a default style. The configuration file option *tangent_edge_display* can be used to set a default tangent display style. The following tangent edge styles are available within Pro/ENGINEER.

- **Solid** Tangent edges are displayed as a solid line.
- **No Display** Tangent edges are not displayed.
- **Phantom** Tangent edges are displayed with a phantom line.
- **Centerline** Tangent edges are displayed with a centerline.
- **Dimmed** Tangent edges are displayed in the color set for dimmed entities, menus, and commands.
- **Tan Default** Tangent edges will be displayed as set in the Environment dialog box.

To change the display mode of an individual view, perform the following steps:

STEP 1: Select VIEWS >> DISP MODE >> VIEW DISP.

STEP 2: Select individual views to modify, then select DONE SEL on the Get Select menu.

STEP 3: Select a Display Mode for the selected views.

Select either Wireframe, Hidden Line, No Hidden, or Default from the View Display menu.

STEP 4: Select a Tangent Edge display style.

From the View Display menu, select a Tangent edge display style.

STEP 5: Select DONE on the View Display menu.

DETAILED VIEWS

Detailed views are used on drawings to highlight portions of a component. A component might have a section with small and complicated features that would be hard to detail within the scope of the drawing's scale. Other sections of a drawing might require a detailed view due to a specific importance factor. Figure 6–7 shows an example of a drawing with a detailed view.

Within Pro/ENGINEER, a Detailed view can be created any time after a general view has been placed. The scale of a detailed view is independent of its parent view. Additionally, Detailed views do not lie along normal lines of projection. By default, a detailed view reflects its parent view. The display mode of a detailed view (line and tangent display) is the same as its parent view. Any cross sections shown in a parent view will be reflected in its detailed views. This view relationship can be broken with the Views >> View Disp >> Det Indep option.

Perform the following steps to create a detailed view:

STEP 1: Select VIEWS >> ADD VIEW.

STEP 2: Select DETAILED >> DONE.

STEP 3: Select the location for the view.

On the work screen, select the location for the detailed view.

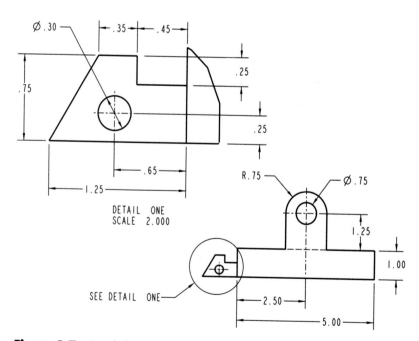

Figure 6-7 Detailed view

STEP 4: **Enter a scale factor for the detailed view.**

If the scale factor is not appropriate, it can be changed at a later time with the Edit >> Value option.

STEP 5: **Select a reference point on the edge of an entity in the parent view.**

Select the entity in the view that includes the portion of the drawing used to create the detailed view. You have to select the edge of an entity. Pro/ENGINEER uses this reference point to regenerate the detailed view.

STEP 6: **Draw a Spline around the geometry to include in the detailed view.**

Use the left mouse button to select points on the spline. Use the middle mouse button to close the spline. Objects included inside the spline boundary will be included in the detailed view.

STEP 7: **Enter a Name for the detailed view.**

As shown in Figure 6–7, Pro/ENGINEER will label the detailed view with the provided name.

STEP 8: **Select a Boundary Type.**

On the parent view, Pro/ENGINEER will enclose the detailed portion in a boundary. There are four boundary types available: Circle, Ellipse, H/V Ellipse, and Spline. For Ellipse, H/V Ellipse, and Spline types, you have to sketch the boundary.

STEP 9: **Place the detailed view's name label.**

For detailed views with a circle as the boundary type, the view will be complete at this point. For Ellipse, H/V Ellipse, and Spline boundary types, you have to select a leader attachment point.

STEP 10: **Select the leader attachment point (Ellipse, H/V Ellipse, and Spline boundary types only).**

SHOWING AND ERASING ITEMS

Drawing mode is used to create annotated presentations of models created in Part and Assembly modes. One of the uses of Drawing mode is to create orthographic views of a model. Orthographic drawings frequently consist of items such as dimensions, centerlines, notes, and geometric tolerances. Within Part mode, features are fully defined through the use of dimensions, constraints, and references. Dimensions used to define a part are considered parametric. When a feature is created, the dimensioning scheme defining the feature should match the intent for the design of the part. These parametric dimensions can be used in Drawing mode to annotate the drawing of the model.

Other items created in Part or Assembly mode can be used in Drawing mode. Drawing mode's Show/Erase option is used to show an item created in Part or Assembly mode or used to erase an item created in Part or Assembly mode. The Erase selection within the Show/Erase option can be misleading. Since the Show/Erase option is used to manipulate the display of items created in Part or Assembly mode, these items cannot be deleted in Drawing mode. A more descriptive name for the Show/Erase option would be Show/No-Show. Items that are erased are actually hidden from display.

The manipulation of the display of items is accomplished through the Show/Erase dialog box (Figure 6–8). Listed are the items that can be shown or erased. To show or erase an item type, the button associated with the item type has to be selected. Multiple item types can be selected at one time. Several options are available for controlling the items that are displayed:

- **Show All** The Show All option will display all items of a particular type. As an example, if Show All is selected with the Dimension item type, all parametric dimensions defining a model will be displayed.

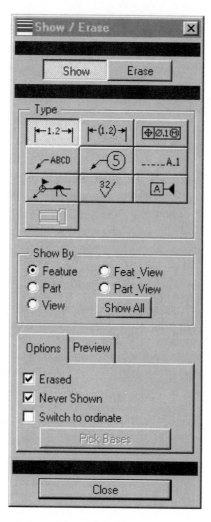

Figure 6-8   Show/Erase dialog box

- **Feature** The Feature option will display items on a selected feature. The user must select the feature on the work screen.

- **Part** The Part option will display items on a selected part. The user must select the part on the work screen or on the model tree. This option is useful for assembly drawings where items on individual parts must be displayed.

- **View** The View option will display items on a selected view. The user must select the view on the work screen.

- **Feat & View** The Feature and View option will display items on a selected feature within the view from which the feature was selected.

- **Part & View** The Part and View option will display items on a selected part within the view from which the part was selected.

- **Erased** The Erased option will show only items that have previously been erased.

- **Never Shown** The Never Shown option will show only items that have never been previously shown.

- **Preview** Located under the Preview tab is the Preview option. The Preview option is used to preview items before they are displayed on the work screen. Options are available for accepting or not accepting an item.

POP-UP MENU

Drawing mode provides a Pop-Up menu for the modification of drawing mode entities to include dimension text and dimension properties. Using the Pop-Up menu, an item can be selected for modification during the drawing process. To modify an item with the Pop-Up menu, select the item with the left mouse button; then with the right mouse button reselect the item. The reselection of the item with the right mouse button will reveal the Pop-Up menu.

DIMENSIONS

Dimensions can be modified in a multiple of ways with the Pop-Up menu. The following dimension modification options are available:

- Dimensions can be moved.
- Dimensions can be switched to another view.
- Dimensions can be jogged.
- Arrows can be flipped.
- Witness lines can be shown or erased.
- Arrow styles can be changed.
- Dimension nominal values can be changed.
- Dimensions can be erased or unerased.
- Dimension values can be changed.

GEOMETRIC TOLERANCES

Geometric tolerances created can be modified or manipulated using the Pop-Up menu. The following options are available:

- Geometric tolerances can be moved on the work screen.
- Geometric tolerance attachments can be modified.
- Geometric tolerances can be switched to another view.
- The leader type of a Geometric tolerance can be changed.
- Geometric tolerances can be redefined.
- Geometric tolerances can be erased or unerased.
- Geometric tolerances can be deleted.

NOTES

Notes can be modified and/or manipulated in the following ways using the Pop-Up menu:

- Notes can be moved on the work screen.
- Notes can be switched to another view.
- Notes text styles can be modified.
- Text can be modified.
- Notes can be erased or unerased.
- Notes can be deleted.

VIEWS

Views can be modified with the Pop-Up menu in the following ways:

- Views may be moved on the work screen.
- The scale of a view can be changed.

- A view's cross section can be replaced.
- The view text can be modified.

DIMENSIONING AND TOLERANCING

Pro/ENGINEER's Drawing mode has the capability to display dimensions created in Part or Assembly mode and the capability to create dimensions within Drawing mode itself. Dimensions created in Part mode can be displayed using the Show/Erase option. These dimensions are associative and can be modified. Due to the associativity between modes, any parametric dimension modified in Drawing mode is also modified in other modes, such as Part and Assembly modes. While parametric dimensions can be hidden with the Show/Erase option, they cannot be deleted. Driven Dimensions can be created within Drawing mode using the Insert >> Dimension option. Dimensions created in this fashion are not associative and are not modifiable. The model geometry drives the value of each dimension.

MANIPULATING DIMENSIONS

Drawing mode provides a variety of options for manipulating dimensions. These tools are used to create a clear and readable engineering drawing.

 MOVE

Drawing mode provides two tools for moving drawing entities. For the manipulation of dimensions and views, the Move icon is available. An alternative to the Move icon is the dragging of drawing entities after their selection. When a dimension is selected, grip points are provided that allow an entire dimension to be repositioned including the dimension's text, dimension line, and witness lines. Additional grip points are provided that allow for the moving of only the dimensions number value.

 SWITCH VIEW

The Switch View icon is used to switch the view in which a dimension is located (Figure 6–9). An alternative to the Switch View icon is the Switch View option available on the pop-up menu. The Switch View option is used often after the Show/Erase >> Show All command. When showing all parametric dimensions in a drawing, Pro/ENGINEER will randomly place dimensions in a view. The placement view may not be the appropriate view according to drafting dimensioning standards.

FLIP ARROW

The Flip Arrow option is used to change the direction that an arrowhead points. This option can be used with linear and radial dimensions (Figure 6–9). The Flip Arrow option is accessible through the pop-up menu by preselecting the dimension text with the left mouse button, followed by reselecting the dimension text with the right mouse button.

MAKE JOG

Dimensions can become too confined when annotating small features. On a normal dimension, the distance between two witness lines is the same as the nominal size of the dimension. The Make Jog option, available by selecting Insert >> Job, can be used to create a jog in a witness line (Figure 6–10).

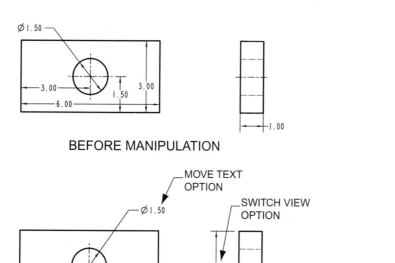

BEFORE MANIPULATION

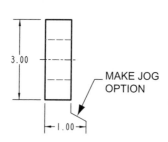

AFTER MANIPULATION

Figure 6-9   Manipulation tools

Figure 6-10   Make Jog option

DIMENSION TOLERANCES AND MODIFICATION

Pro/ENGINEER allows for the creation of dimensions that adhere to ANSI or ISO tolerance standards. See Chapter 2 for more information on setting a tolerance standard. Within Drawing mode, tolerances may be displayed in a variety of formats. Available formats include Limits, PlusMinus, and PlusMinusSymmetric. In Drawing mode, before a dimension can be displayed as a tolerance, the Drawing Setup file option *Tol_Display* has to be set to Yes. Use the Advanced >> Draw Setup option and set the *Tol_Display* option to modify the tolerance display for an individual drawing.

When utilizing ANSI as the tolerance standard, tolerance values and formats can be set with the Properties option on Drawings mode's Pop-Up menu, which displays the Modify Dimension dialog box (Figure 6–11). The Modify Dimension dialog box is used to set the following dimension options:

- **Tolerance mode**   Nominal, PlusMinus, and PlusMinusSymmetric may be selected.

- **Tolerance values**   Tolerance values associated with a particular format may be entered.

- **Decimal places**   The number of decimal places of a dimension may be selected.

- **Basic dimension**   When utilizing geometric tolerances, a dimension may be set as basic.

- **dimension text**   The Dimension Text tab (Figure 6–11) is used to add text and symbols around a dimension value. When adding text, the dimension value cannot be deleted. The symbol pallet is available for the selection of symbols (Figure 6–12). Figure 6–13 shows an example of how a typical dimension text note can be changed into a counterbored hole note.

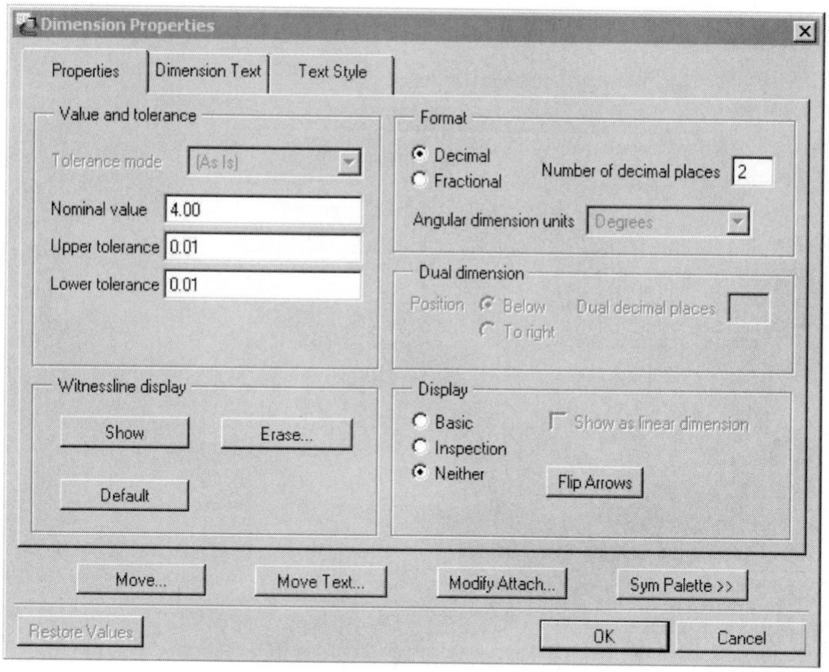

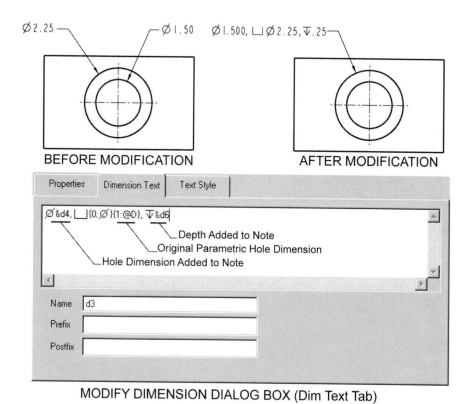

Figure 6-11   Modify Dimension dialog box

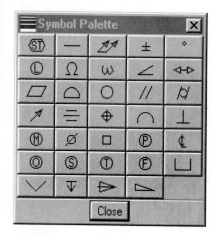

Figure 6-12   Symbol pallet

Ø 2.25 — ⊙ — Ø 1.50   Ø 1.500, ⊔ Ø 2.25, ▽.25 —

BEFORE MODIFICATION   AFTER MODIFICATION

| Properties | Dimension Text | Text Style |

Ø &d4, ⊔ {0: Ø }{1:@D}, ▽ &d6

Depth Added to Note
Original Parametric Hole Dimension
Hole Dimension Added to Note

Name   d3

Prefix

Postfix

MODIFY DIMENSION DIALOG BOX (Dim Text Tab)

Figure 6-13   Dimension text modification

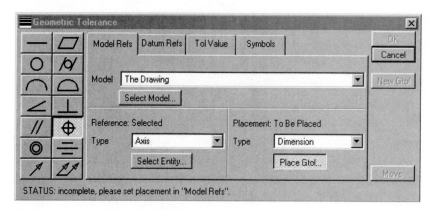

Figure 6-14   Geometric Tolerance dialog box

> **MODELING POINT**   Notice in Figure 6–13 the 1.50 diameter hole as displayed in the drawing before modification. This value is a parametric dimension used to define the size of the hole. This dimension can be modified, which after regeneration will modify the part's hole diameter. By entering the hole's assigned dimension symbol in the Dim Text tab (e.g., *&d11*) instead of its dimension value, the "after modification" hole note can be used to modify its corresponding parametric dimension.

GEOMETRIC TOLERANCES

Geometric tolerances are used to control geometric form, orientation, and location. An example of a geometric tolerance would be specifying that a planar surface is flat to within a tolerance value. The Geometric Tolerance dialog box is used to create a geometric tolerance characteristic (Figure 6–14). This dialog box can be accessed through the Insert >> Geometric Tolerance menu option. See Chapter 1 for more information on establishing a geometric tolerance. Before a Datum can be utilized in a drawing, it has to be first set through the Edit >> Properties option on Pro/ENGINEER's menu bar.

SUMMARY

Drawing mode is used to create drawings of Pro/ENGINEER parts or assemblies. Multiple views of a model can be displayed to include projection views, section views, and partial views. Dimensions and other items such as cosmetic threads and geometric tolerance notes can be displayed in a drawing using the Show/Erase option. Drawing mode has the capability for the creation of nonassociative two-dimensional drawings. Notes, leaders, and nonparametric dimensions can be added to a drawing.

DRAWING TUTORIAL 1

This tutorial will demonstration the creation of the drawing shown in Figure 6–15. The part used in this tutorial is shown in Figure 6–16. The start of this tutorial will require you to model this part. As shown in Figure 6–15, four different view types will be created: General, Projection, Auxiliary, and Detailed.

This tutorial will cover:

- Starting a drawing.
- Adding a drawing format.
- Creating a general view.
- Creating projection views.
- Creating a detail view.
- Creating notes.
- Modifying the drawing setup file.

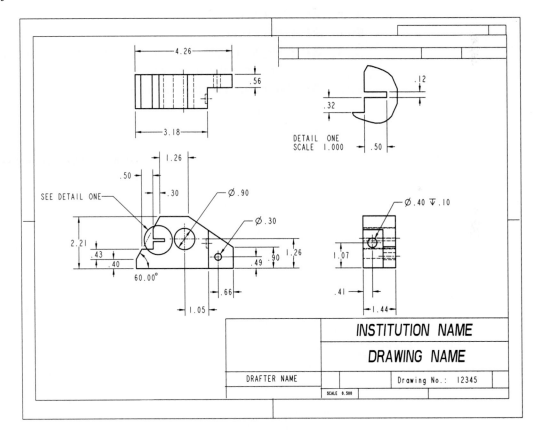

Figure 6-15   Multiview drawing

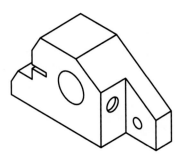

Figure 6-16   Model for drawing

CREATING THE PART

Model the part shown in Figures 6–15 and 6–16, naming it *view1*. The dimensioning scheme shown in Figure 6–15 matches the design intent for the part. When modeling this part, make sure that these dimensions are incorporated into your design.

STARTING A DRAWING

This section of the tutorial will create the object file for the drawing to be completed. Before starting this tutorial, ensure that you have completed the part model shown in Figure 6–16.

STEP   1: Start Pro/ENGINEER.

If Pro/ENGINEER is not open, start the application.

STEP   2: Set an appropriate Working Directory.

STEP   3: Select FILE >> NEW.

STEP   4: In the New dialog box, select DRAWING mode then enter *VIEW1* as the name of the drawing file (Figure 6–17).

STEP   5: Select OKAY on the New dialog box.

After selecting OKAY on the dialog box, Pro/ENGINEER will reveal the New Drawing dialog box. This dialog box is used to select a model, a sheet size, and a format.

STEP   6: Select the BROWSE option on the New Drawing dialog box and locate the *VIEW1* part (Figure 6–18).

Use the Browse option to locate the *view1* part created in the first section of this tutorial. The part will serve as the Default Model for the creation of drawing views. If *view1* is the active part during the creation of this drawing file, it will be displayed by default in the model selection box.

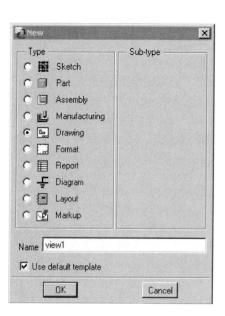

Figure 6-17   New dialog box

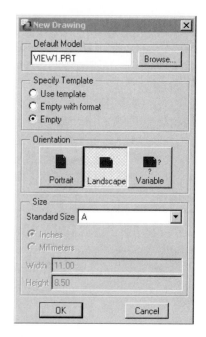

Figure 6-18   New Drawing dialog box

STEP 7: **Select the EMPTY option (Figure 6–18) under the Specify Template option.**

The Specify Sheet option allows for the selecting of a standard or user-defined sheet size. The Retrieve Format option allows for the retrieval of a predefined sheet format. When retrieving a format, the size of the sheet defining the format will define the sheet size for the new drawing.

STEP 8: **Select LANDSCAPE as the Orientation option.**

STEP 9: **Select A as the Standard Sheet Size (Figure 6–18).**

Pro/ENGINEER provides a variety of standard sheet sizes (A, B, C, A1, A2, etc.). A unique sheet size can be entered in either inch or millimeter units.

STEP 10: **Select OKAY on the New Drawing dialog box.**

After selecting OKAY, Pro/ENGINEER will launch its Drawing mode.

ADDING A DRAWING FORMAT

This section of the tutorial will provide instruction on how to add a format to a drawing sheet. Formats can be retrieved from a variety of sources. Pro/ENGINEER provides sheet formats in the *format* subdirectory of the Pro/ENGINEER load point directory. Formats can also be imported through IGES and DXF. The configuration file option *pro_format_dir* is used to specify a default format directory.

STEP 1: **Select SHEETS >> FORMAT >> ADD/REPLACE.**

On the Menu Manager, select the Sheets menu option; then select the Format option. The Add/Replace option is used to add formats to a drawing sheet or to replace an existing format.

STEP 2: **Locate, select, and open an A size sheet format (Figure 6–19).**

A format added to a drawing is associated with its original format file. Any changes made to the format through Format mode will be reflected in all drawings using that specific format.

STEP 3: **Select DONE/RETURN to exit the Sheets menu.**

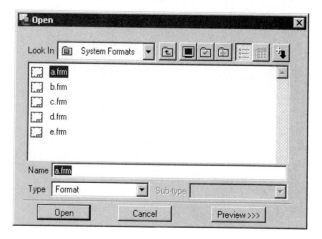

Figure 6-19 Opening an A size format

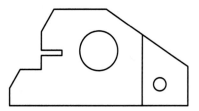

Figure 6-20   Front view of the part

CREATING A GENERAL VIEW

General views serve as the parent view for all views projected off of it. This section of the tutorial will create a General view as the Front view of the drawing (Figure 6–20).

STEP   1:   Select the VIEWS menu option.

STEP   2:   Select ADD VIEW >> GENERAL >> FULL VIEW on the Views menu.

The Add View option is selected by default. Since no views currently exist within the drawing, General is the only view type available.

STEP   3:   Select NoXsec >> SCALE on the View Type menu.

No Cross Section (No Xsec) is selected by default. The Section option will allow for the creation or retrieval of a section view. The Scale option allows for the entering of a scale value for the drawing view.

STEP   4:   Select DONE on the View Type menu.

STEP   5:   On the work screen, select the location for the drawing view.

The drawing under creation in this tutorial will consist of a Front, Right-side, Top, and detailed view. The general view currently being defined will serve as the Front view. On the work screen, pick approximately where the front view will be located. The Views >> Move View option can be used to reposition this view's location.

STEP   6:   In the textbox, enter .500 as the scale value for the view.

A scale value of 0.500 will create a view at half scale. The defining of a scale value is required due to the selection of the Scale option in step 3.

After entering the scale value, Pro/ENGINEER will insert the model onto the work screen with the default orientation. The Orientation dialog box will be used to orient the view correctly.

STEP   7:   On the Orientation dialog box, select FRONT as the Reference 1 option.

> **MODELING POINT**   When selecting references to orient a model, it is often helpful to turn off all datum planes and to display the model as No Hidden (Figure 6–21). This technique provides clarity when selecting a reference surface.

A planar surface selected with the Front reference option will orient the surface toward the front of the work screen. Other available reference options include Back, Top, Bottom, Left, Right, Vertical Axis, and Horizontal Axis.

Figure 6-21   No hidden display and datum planes off

REFERENCE 2: TOP

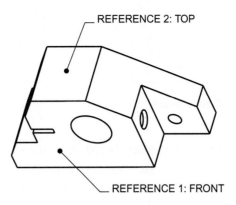

REFERENCE 1: FRONT

Figure 6-22   Orienting the model

STEP 8: **Pick the front of the model (Figure 6–22).**

STEP 9: **On the Orientation Dialog box, select TOP as the Reference 2 option.**

> A planar surface selected with the Top reference option will orient the surface toward the top of the work screen.

STEP 10: **Select the top of the model (Figure 6–22).**

STEP 11: **Select OKAY on the orientation dialog box.**

> Your front view should appear as shown in Figure 6–20.

CREATING PROJECTION VIEWS

Once a General view has been created, views can be projected off of it. Options exist for creating Projection, Auxiliary, and/or Section views. This segment of the tutorial will create a right-side view and a top view, both projected from the front view (Figure 6–23). The front view will be the parent view of these projected views.

STEP 1: **Select VIEWS >> ADD VIEW.**

STEP 2: **Select PROJECTION >> FULL VIEW on the View Type menu.**

> The Projection option will project a view from an existing view. In this step of the tutorial, the right-side view will be projected from the front view.

STEP 3: **Select NoXsec >> NO SCALE on the View Type menu.**

> Since this view will be projected from an existing view, it will take the scale value of its parent view.

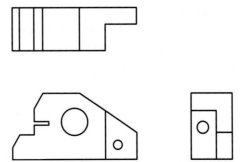

Figure 6-23   Front, top, and right-side views

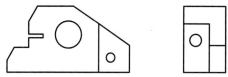

Figure 6-24   Front and right-side views

STEP   4:   **Select DONE from the View Type menu.**

Did you notice how steps 2 and 3 required the selection of default options? After a general view has been added to a drawing, Pro/ENGINEER defaults to the Projection >> Full View >> NoXsec >> No Scale options.

STEP   5:   **On the work screen, select the location for the Right-Side view.**

Pro/ENGINEER will create a projected view based upon its parent view. Your view should appear as shown in Figure 6–24. The Views >> Move View option can be used to reposition a view.

STEP   6:   **Select VIEWS >> ADD VIEW.**

Next, you will add the Top view.

STEP   7:   **Select DONE on the View Type menu.**

You will take the default options found under the View Type menu (Projection >> Full View >> NoXsec >> No Scale).

STEP   8:   **On the work screen, select the location for the top view.**

Your views should appear as shown in Figure 6–23.

STEP   9:   **Use the VIEWS >> MOVE VIEW option to reposition each view on the work screen.**

When moving a general view, all views projected from it will be repositioned to keep normal lines of projection. When moving a projected view, it will remain aligned with its parent view. Views can also be moved by dragging them on the screen with the left mouse button.

STEP 10:   **If necessary, select DONE/RETURN to exit the Views menu.**

STEP 11:   **Save your drawing**

CREATING A DETAILED VIEW

Pro/ENGINEER's Drawing mode has an option for creating a detailed view. Detailed views are used to highlight and expand an area of an object that requires special attention. An example would be a small, complex feature that would be hard to dimension on a normal projection view. This segment of the tutorial will create the detailed view shown in Figure 6–25.

STEP   1:   **Select VIEWS >> ADD VIEW.**

STEP   2:   **Select DETAILED as the view type to add.**

STEP   3:   **Select DONE on the View Type menu.**

STEP   4:   **On the work screen, select the location for the detailed view (Figure 6–25).**

Using the mouse, select on the work screen where the detailed view will be located.

STEP   5:   **Enter 1.00 as the scale value for the detailed view.**

This step will create a detailed view that is full scale.

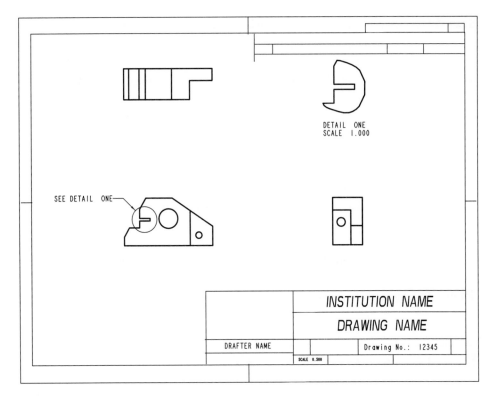

Figure 6-25   Detailed view of part

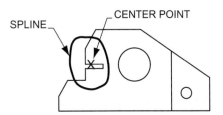

Figure 6-26   Sketching the spline

Step   6:   On the front view, pick an entity centered within the detail area (Figure 6–26).

Pro/ENGINEER requires the selection of an entity in an existing view. This selection point is used to calculate the regeneration of the detailed view.

Step   7:   As shown in Figure 6–26, sketch a spline around the area to detail.

Using the mouse, sketch a spline that will include the area to be detailed. Use the left mouse button to select spline points and the middle mouse button to close the spline.

Step   8:   Enter *ONE* as the name for the detailed view.

Step   9:   Select CIRCLE as the Detailed view boundary type.

The Circle option will include the area to be detailed in a circle (see Figure 6–25). Other options available include Ellipse, H/V Ellipse, and Spline.

STEP 10: **Select a location on the work screen for the Detail Note.**

As shown previously in Figure 6–25, this will locate the note *SEE DETAIL ONE*. This note can be moved later if necessary.

STEP 11: **Use the VIEWS >> MOVE VIEW option to reposition the detailed view.**

If necessary, reposition the detailed view or any other view.

STEP 12: **Select DONE/RETURN on the Views menu.**

Within Pro/ENGINEER, it is important for menu management to properly exit a menu. Always use an available Done or Done/Return option to properly exit any menu.

STEP 13: **If necessary, select the detail view's note and drag it to an appropriate location.**

Views, notes, dimensions, and other annotations can be moved by selecting the entity with the left mouse button. Your drawing should appear as shown in Figure 6–25.

STEP 14: **Save your drawing.**

ESTABLISHING DRAWING SETUP VALUES

Pro/ENGINEER's drawing setup file is used to set option values associated with a drawing. Examples of options that can be set include text height, arrowhead size, and geometric datum plane symbol. Multiple drawing setup files can be created. This segment of the tutorial will modify the values of the current drawing setup file.

STEP 1: **Select ADVANCED >> DRAW SETUP.**

After selecting Draw Setup, Pro/ENGINEER will open the active drawing settings within an Options dialog box. This dialog box is similar to the Preferences dialog box used to make changes to the configuration file option.

STEP 2: **Change the Text and Arrowhead values for the current drawing.**

Within the Options dialog box, make changes to the options shown in Table 6–2. To change an option's value, perform the following steps:

1. Type the option's name in the Option field.
2. Type or select the option's value in the Value field.
3. Select the Add/Change icon.

STEP 3: **Save the modified values for the active drawing setup file, then Exit the text editor.**

Table 6-2 Values for drawing setup

Drawing Setup Option	New Value
drawing_text_height	0.125
text_width_factor	0.750
dim_leader_length	0.175
dim_text_gap	0.125
draw_arrow_length	0.125
draw_arrow_style	filled
draw_arrow_width	0.0416

CREATING DIMENSIONS

Pro/ENGINEER provides options for creating two types of dimensions in Drawing mode: parametric and non-parametric. The Show/Erase dialog box can be used to show parametric dimensions that were created in Part and Assembly modes. During part modeling, a sketched feature has to be fully defined by utilizing dimensions, constraints, and references. The dimensions defining a feature can be revealed in Drawing mode through the Show and Erase option. The second option, Insert >> Dimension, can be used to create non-parametric dimensions.

This segment of the tutorial will create the dimension annotations shown in Figure 6–27.

STEP 1: **On the menu bar, select VIEW >> SHOW AND ERASE.**

The Show and Erase option is used to show and not show items that were defined in Part and/or Assembly modes. Figure 6–28 shows the Show/Erase dialog box. Items that can be shown include parametric dimensions, reference dimensions, geometric tolerances, notes, balloon notes, axes (centerlines), symbols, surface finish, datum planes, and cosmetic features.

STEP 2: **On the Show/Erase dialog box, select the Dimension item type (Figure 6–29).**

The dimension item type option will show parametric dimensions that were created in Part and Assembly modes.

STEP 3: **On the Show/Erase dialog box, under the Options tab, check ERASED and NEVER SHOWN options (Figure 6–28).**

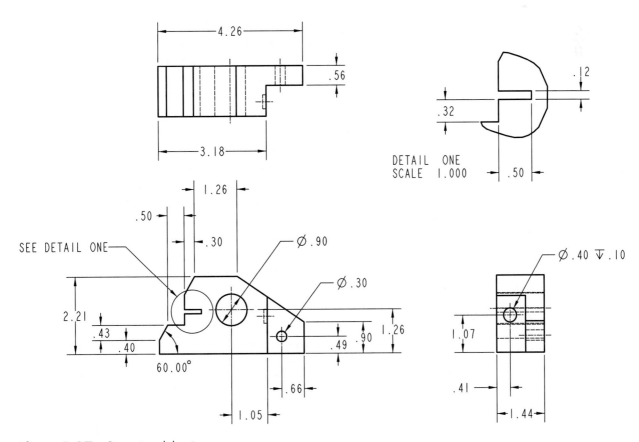

Figure 6-27   Dimensioned drawing

Figure 6-28   Show/Erase
dialog box

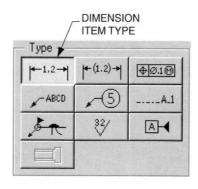

Figure 6-29   Dimension item type

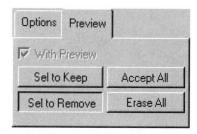

Figure 6-30   With Preview option

The Erased option will show items that have previously been erased, and the Never Shown option will show items that have never been shown.

STEP 4: On the Show/Erase dialog box, under the Preview tab, check the WITH PREVIEW option (Figure 6–30).

STEP 5: Select SHOW ALL (Figure 6–28) on the Show/Erase dialog box and Confirm the selection.

The Show All option will show all available item types. In this step of the tutorial, all dimensions available from the referenced model will be shown. As revealed in Figure 6–31, this can create a confusing and cluttered drawing. You will use the Move, Switch View, Flip Arrows, and Erase options to clean the drawing.

STEP 6: Select ACCEPT ALL under the Preview option (Figure 6–30).

The previously selected With Preview option allows for the previewing of shown item types. Use the Accept All option to accept the shown dimensions.

STEP 7: CLOSE the Show/Erase dialog box.

STEP 8: Use Drawing mode's MOVE and MOVE TEXT capabilities and the Pop-Up menu's SWITCH VIEW and FLIP ARROWS options to reposition the dimensions to match Figure 6–32.

The Pop-Up menu is available by first selecting the text of a dimension with your left mouse button, followed by reselecting the same dimension text with your right mouse button (*Note:* A slight delay in releasing the right mouse button is necessary). A variety of modification and manipulation tools are available through the revealed Pop-Up menu.

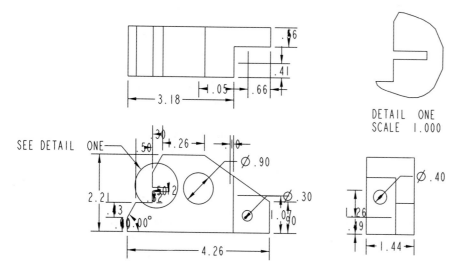

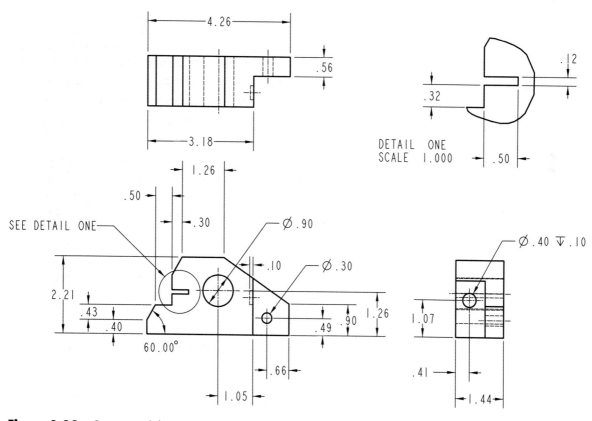

Figure 6-31 Dimensions shown on a drawing

Figure 6-32 Repositioned dimensions

Use the following options to reposition your dimensions on the work screen:

• **Move** The Move option is available through the selection of a dimension's text with the left mouse button. Once selected, the dimension, including text and witness lines, can be dragged with your cursor.

- **Move Text** The Move Text option is available through the selection of a dimension's text with the left mouse button. Once selected, grip points are available that allow you to move only the text of a dimension.

- **Switch View** The Switch View option is used to switch a dimension from one view to another. It is available under the Pop-Up menu. Multiple dimensions can be selected by a combination of the Shift key and the left mouse button. Once dimensions are preselected, the Pop-Up menu is available through the right mouse button.

- **Flip Arrows** The Flip Arrows option is used to flip dimension arrowheads. It is available under the Pop-Up menu.

> **MODELING POINT**   The Switch View icon is also available on the Drawing toolbar. It is a handy option for switching entities, to include dimension lines, from one view to a second view.

STEP 9: **Use the SHOW/ERASE option to show all centerlines.**

The Axis option (Figure 6–33) on the Show/Erase dialog box is used to show centerlines. On the Show/Erase dialog box, select the Axis option then select Show All. Accept all available centerlines then close the dialog box.

STEP 10: **Use the left mouse button's move capability to create witness line gaps.**

Where a dimension's witness line is linked to an entity (Figure 6–34), a visible gap (approx. 1/16-inch wide) is required. After selecting the dimension's text with the left mouse button, drag the witness line's grip point to create the gap.

STEP 11: **Pick the .50-inch diameter hole dimension; then select the Pop-Up menu's PROPERTIES option.**

STEP 12: **Use the Dimension Text tab on the dimension Properties dialog box to modify the hole's note (Figure 6–35).**

You will add the .10-inch depth value to the .40-inch diameter hole note. Under the Dimension Text tab, use the Sym Palette and the keyboard to add the hole depth attribute shown in the illustration.

STEP 13: **Select the VIEW >> SHOW AND ERASE option.**

The 0.10 dimension shown in Figure 6–36 is not needed in the drawing. The Show/Erase option can be used to remove it from the drawing.

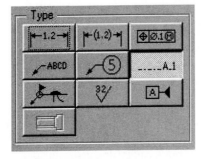

Figure 6-33   Axis item type

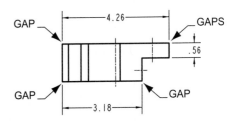

Figure 6-34   Witness line gaps

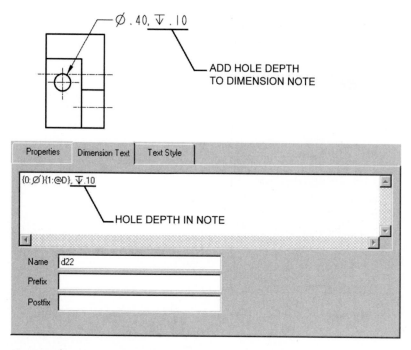

Figure 6-35   Add text and symbols to a dimension note

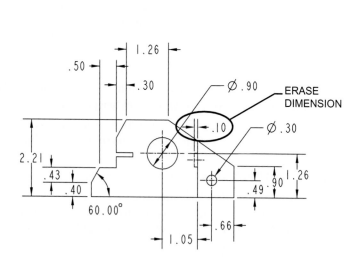

Figure 6-36   Erase dimension

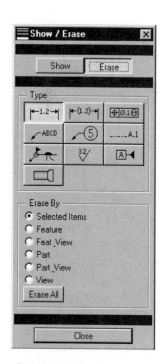

Figure 6-37   Dimension item type

STEP 14: On the Show/Erase dialog box, select the ERASE option, then select the DIMENSION item type (Figure 6–37).

STEP 15: On the work screen, select the .10 dimension text; then select the middle mouse button.

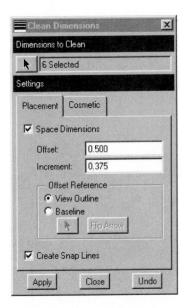

Figure 6-38   Clean Dimensions
dialog box

The Erase option does not delete a selected item. Instead, it removes it from the drawing screen. Any erased item can be redisplayed with the Show option.

STEP 16: Select the **CLEAN DIMENSIONS** icon on the **Drawing toolbar.**

The Clean Dimensions option will consistently space dimensions on a drawing view. The minimum distance between a dimension line and the object is 3/8 inch, while the minimum spacing between two dimension lines is 1/4 inch. These values can be increased when appropriate. After selecting the Clean Dims option, the Clean Dimensions dialog box will appear (Figure 6–38).

STEP 17: On the work screen, select the front, top, right-side, and detailed views, then select **DONE SEL** on the **Get Select** menu.

Pro/ENGINEER's default values for the Clean Dimensions dialog box are 0.500 for the object Offset setting and 0.375 for the dimension Increment setting. You will keep these default settings.

STEP 18: Select **APPLY,** then **CLOSE** the dialog box.

After applying the clean dimension settings, Pro/ENGINEER will create snap lines that will evenly space any dimension line. Depending upon the complexity of the drawing, the applied settings might not create an ideal drawing. Snap lines will not print when the drawing is plotted.

STEP 19: Use Drawing modification and manipulation tools to refine the placement of dimensions.

Use available tools such as Move View, Move, Flip Arrows, and Move Text to tweak the placement of dimensions. Your final dimensioning scheme should appear as shown in Figure 6–39.

STEP 20: Save your drawing.

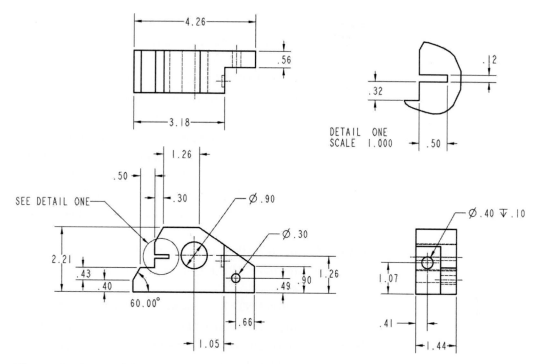

Figure 6-39   Final dimensioning scheme

CREATING NOTES

Annotations can be added to a drawing using the Insert >> Note option. Within the Note Types menu, options are available for creating notes with or without leader lines. Also, options are available for justifying text and for entering text from a file. This segment of the tutorial will create text for the title block. The finished title block will appear as shown in Figure 6–40.

STEP   1:   **Using the Pop-Up menu's Modify Text Style option, modify the *SCALE 0.500* note to have a text height of 0.100 inches.**

When a General view is added to a drawing, Pro/ENGINEER inserts a scale value. In this tutorial, the first general view has a scale value equal to 0.500. When selecting the SCALE 0.500 note, you have to select both the word *SCALE* and the number *0.500*.

STEP   2:   **Move the *SCALE 0.500* note to the title block (Figure 6–40).**

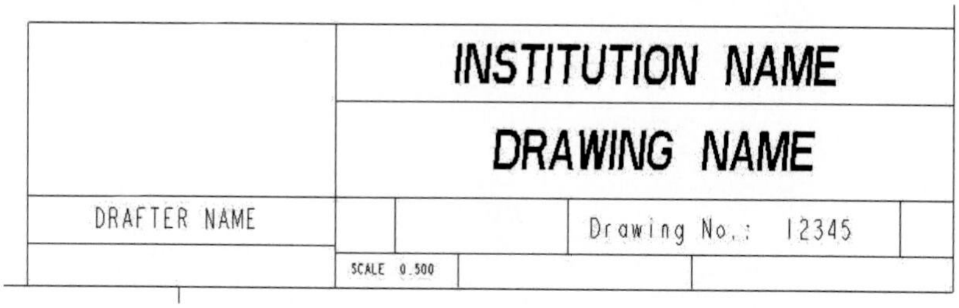

```
┌─────────────┬──────────────────────────────┐
│             │     INSTITUTION NAME         │
│             ├──────────────────────────────┤
│             │      DRAWING NAME            │
├──────┬──────┼────────────┬─────────────────┤
│ DRAFTER NAME│            │ Drawing No.: 12345│
├──────┴──────┼────────────┼─────────────────┤
│      │SCALE 0.500│        │                 │
└──────┴──────┴────────────┴─────────────────┘
```

Figure 6-40   Title block information

STEP 3: **Select INSERT >> NOTE and create the text shown in the figure.**

The Insert >> Note option is used to add text to a drawing. Use the following options to create each note:

- **No leader** Creates a note without a leader line.
- **Enter** Allows for the creation of a note through the keyboard.
- **Horizontal** Creates notes horizontal on the work screen.

STEP 4: **Select CENTER on the Note Types menu.**

The Center option will center justify the note text. Other justification options available include Left, Right, and Default.

STEP 5: **Select MAKE NOTE to enter the text for the note.**

STEP 6: **Within the work screen, select the location for the *INSTITUTION NAME* note.**

STEP 7: **In the textbox, enter your institution's name as the note text.**

In Pro/ENGINEER's textbox, enter the text; then select the Enter key on the keyboard.

STEP 8: **Select ENTER on the keyboard to end the Make Note option.**

STEP 9: **Repeat the note making steps to create the remaining notes.**

STEP 10: **Select DONE/RETURN from the Note Types menu to exit the note creation menu.**

STEP 11: **On the work screen, using a combination of the Shift key and the left mouse button, preselect the *INSTITUTION NAME* and the *DRAWING NAME* notes.**

STEP 12: **Access the Pop-Up menu's Modify Text Style option.**

You will modify the text style and the text height of the *Institution Name* and *Drawing Name* notes.

STEP 13: **On the Text Style dialog box, select FILLED as the Font (Figure 6–41).**

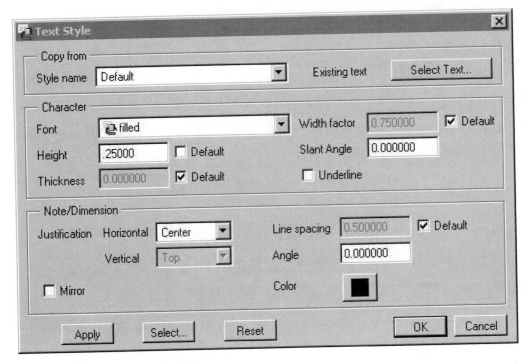

Figure 6-41   Text Style dialog box

Step 14: On the Text Style dialog box, enter a text Height of 0.250.

Step 15: Select OKAY on the Text Style dialog box.

Step 16: Save your drawing.

SETTING DISPLAY MODES

The display of hidden lines on a drawing can be controlled with the display model options found on the toolbar (e.g., Wireframe, Hidden Line, and No Hidden). Additionally, the tangent edge display mode can be set with the Environment dialog box. The problem with using these two methods to set the display mode of a drawing is that a selected setting, such as hidden line display, will affect the entire drawing. Often, it is necessary to set a different display mode for a view or for an individual entity. The following steps will set the front, top, and right side views with a Hidden Line display. Additionally, the detailed view will be set with a No Hidden display.

Step 1: Select VIEWS >> DISP MODE >> VIEW DISP.

Step 2: On the work screen, select the front, top, and right-side views.

Step 3: Select DONE SEL on the Get Select menu (or select the middle mouse button).

Step 4: Select HIDDEN LINE as the display mode.

Other options available include Wireframe, No Hidden, and Default. The Default setting assumes the display mode selected from the toolbar.

Step 5: Select DONE on the View Display menu.

Step 6: Select VIEWS >>DISP MODE >> VIEW DISP.

You will next set a No Hidden display mode for the Detailed view.

Step 7: On the work screen, select the Detailed view; then select DONE SEL.

Step 8: Select DET INDEP on the View Display.

The Detail Independent (Det Indep) selection makes the Detail's display mode independent from its parent's display mode.

Step 9: Select NO HIDDEN as the display mode.

Step 10: Select DONE on the View Display menu.

Step 11: Save your drawing and purge old versions of the drawing by using the FILE >> DELETE >> OLD VERSIONS option.

DRAWING TUTORIAL 2

This tutorial will create the drawing shown in Figure 6–42. As with the first tutorial in this chapter, the start of this tutorial will require you to model the part. As shown in Figure 6–42, two different view types will be created: a General view and a Projection view.

As shown in Figure 6–42, this tutorial will demonstrate the creation of tolerances on a Pro/ENGINEER drawing. This tutorial will cover:

- Starting a drawing with a drawing format.
- Creating a general view.
- Creating a projection view.
- Adding geometric tolerances to a drawing.
- Setting dimensional tolerances.
- Modifying the drawing setup file.

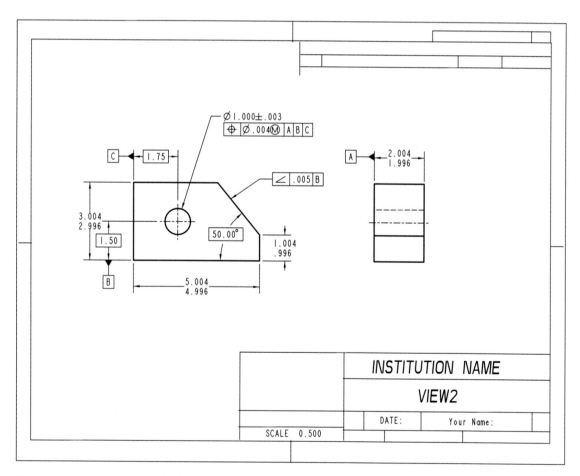

Figure 6-42   Pro/ENGINEER drawing with geometric tolerances

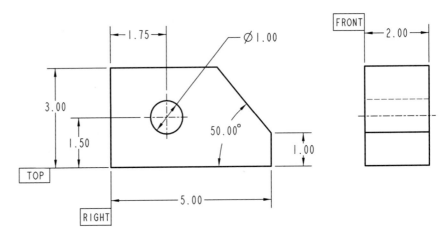

Figure 6-43   Part and dimensioning scheme

CREATING THE PART

Model the part shown in Figure 6–43, naming the part *VIEW2*. When modeling the part, make sure that the locations of your datum planes match the locations of the datum planes shown in the drawing. As a reference, later in this tutorial, the default datum planes RIGHT, TOP, and FRONT will be renamed A, B, and C respectively. The required dimensioning design intent is shown. When modeling this part, make sure that these dimensions are incorporated into your design.

STARTING A DRAWING WITH A TEMPLATE

This section of the tutorial will create the object file for the drawing to be completed. You will utilize an existing drawing template file to set views, formats, and other drawing settings. Before starting this segment of the tutorial, ensure that the part shown in Figure 6–43 has been completed.

STEP   1:   Start Pro/ENGINEER.

If Pro/ENGINEER is not open, start the application.

STEP   2:   Set an appropriate Working Directory.

STEP   3:   Select FILE >> NEW (Use the Default Template option).

STEP   4:   In the New dialog box, select Drawing mode; then enter *VIEW2* as the name of the drawing file.

STEP   5:   Select OKAY on the New dialog box.

After selecting OKAY on the dialog box, Pro/ENGINEER will reveal the New Drawing dialog box. This dialog box is used to select a model to create the drawing from and to set a sheet size.

STEP   6:   On the New Drawing dialog box, select the BROWSE option and locate the *VIEW2* part (Figure 6–44).

Use the Browse option to locate the *VIEW2* part created in the first section of this tutorial. The part will serve as the Default Model for the creation of the drawing views.

STEP   7:   On the New Drawing dialog box, under the Specify Template option, select the EMPTY WITH FORMAT option (Figure 6–44).

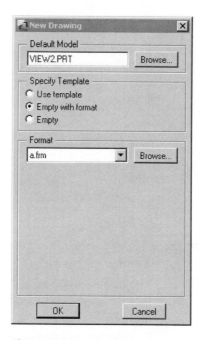

Figure 6-44   New Drawing dialog box

The Specify Sheet option allows for the selection of a standard or user-defined sheet size. The Empty-with-Format suboption allows for the retrieval of a predefined sheet format. When retrieving a format, the size of the sheet defining the format will define the sheet size for the new drawing.

Template files under the Use Template option come with preestablished settings such as views and model display properties. As an example, the *a_drawing* template comes complete with preexisting front, top, and right side views. Other information that can preexist in a template file includes drawing notes not derived from drawings model (e.g., notes and symbols) and parameter notes defined within the model.

STEP   8:   **Select the BROWSE option under the Format section of the New Drawing dialog box.**

The second Browse option on the dialog box allows you to browse the directory structure to locate existing sheet formats.

STEP   9:   **Open a.frm as the standard sheet format.**

Pro/ENGINEER provides a variety of standard sheet formats (e.g., A, B, C, D, and E). A user-defined format can also be selected. When picking an existing format, the size of the format defines the size of the drawing sheet.

STEP 10:   **Select OKAY on the New Drawing dialog box.**

ESTABLISHING DRAWING SETUP VALUES

This section of the tutorial will temporarily set Pro/ENGINEER's drawing settings. The Advanced >> Draw Setup option will be used to change the default settings for this specific drawing.

STEP 1:   **Select ADVANCED >> DRAW SETUP.**

After selecting Draw Setup, Pro/ENGINEER will open the active drawing settings within an Options dialog box. This dialog box is similar to the Options dialog box used to make changes to configuration file options.

Table 6-3   Values for drawing setup

Drawing Setup Option	New Value
drawing_text_height	0.125
text_width_factor	0.750
dim_leader_length	0.175
dim_text_gap	0.125
tol_display	YES
draw_arrow_length	0.125
draw_arrow_style	FILLED
draw_arrow_width	0.0416
gtol_datums	STD_ISO

STEP 2: **Change the Text, Arrowhead, and Datum values for the current drawing.**

Within the Options dialog box, make changes to the options shown in Table 6–3. To change an option's value, perform the following steps:

1. Type the option's name in the Option field.

2. Type or select the option's value in the Value field.

3. Select the Add/Change icon.

The item *tol_display* will display dimensions in tolerance mode. The item *gtol_datums,* when set to a value of *STD_ISO,* will display datums in the ISO format.

STEP 3: **Apply and Close the modified values for the active drawing setup file.**

Save your drawing setup file values, then exit the text editor.

STEP 4: **Select DONE/RETURN to close the Advanced Drawing Options menu.**

CREATING THE GENERAL VIEW

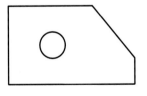

Figure 6-45   Front View

A general view serves as the parent view for all views projected off of it. This section of the tutorial will create a General view as the Front view of the drawing, as shown in Figure 6–45.

STEP 1: **Select VIEWS >> ADD VIEW >> GENERAL >> FULL VIEW.**

The Add View option is selected by default. Since no views currently exist within the drawing, General is the only view type available.

STEP 2: **Select NoXsec >> SCALE >> DONE.**

No-Cross-Section (No Xsec) is selected by default. The Section option allows for the creation or retrieval of a section view. The Scale option allows for the entering of a scale value for the drawing view.

STEP 3: **On the work screen, select the location for the Front view.**

The drawing under creation in this tutorial will consist of a Front view and a Right-side view. The general view currently being defined will serve as the Front view. On the work screen, pick approximately where the Front view will be located. Once placed, the Views >> Move View option can be used to reposition the location of any view.

STEP 4: **Enter .500 as the scale value for the view.**

A scale value of 0.500 will create a view at half scale. The defining of a scale value is required due to the selection of the Scale option in Step 2.

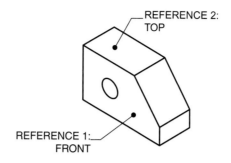

Figure 6-46   Orienting the model

After entering the scale value, Pro/ENGINEER will insert the model into the work screen with the default orientation. The Orientation dialog box will be used to orient the view correctly.

INSTRUCTIONAL NOTE   Depending upon your part construction method, your default orientation might be different from the orientation shown in Figure 6–46. Adjust your view orientation reference selections accordingly. Your final view should match the front view shown in Figure 6–45.

STEP   5: On the Orientation dialog box, select FRONT as the Reference 1 option.

MODELING POINT   When selecting references to orient a model, it is often helpful to turn off all datum planes and to display the model as No Hidden. This technique will provide clarity when selecting a reference surface.

STEP   6: Select the front of the model (Figure 6–46)

A planar surface selected with the Front reference option will orient the surface toward the front of the work screen.

STEP   7: On the Orientation Dialog box, select TOP as the Reference 2 option.

STEP   8: Select the top of the model that will point toward the top of the work screen (Figure 6–46).

Select the surface shown in the figure. If your default model does not appear as shown in the illustration, adjust your reference selection accordingly. Your final orientation should appear as shown in Figure 6–45.

STEP   9: Select OKAY on the orientation dialog box

Your front view should appear as shown in Figure 6–45.

STEP 10: Save your drawing

Drawings are saved with an *.drw* file extension.

CREATING THE RIGHT-SIDE VIEW

This segment of the tutorial will create the right-side view of the part (Figure 6–47).

STEP 1: Select VIEWS >> ADD VIEW.

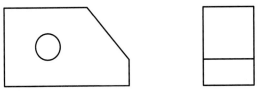

Figure 6-47   Front and right-side views

Figure 6-48   Model Display toolbar

STEP 2:   **Select PROJECTION >> DONE.**

Since Projection, Full View, and NoXsec are the default selections, you only have to select Done to perform this step.

STEP 3:   **On the work screen, select the location for the Right-Side view.**

Pro/ENGINEER will create a projected view based upon its parent view. Your view should appear as shown in Figure 6–47. The Views >> Move View option can be used to reposition a view.

STEP 4:   **On the toolbar, select the Hidden Line display icon (Figure 6–48).**

Selecting Hidden Line as the model display mode will produce hidden lines when the drawing is plotted.

STEP 5:   **If necessary, select DONE/RETURN to exit the Views men.**

SETTING AND RENAMING DATUM PLANES

This section of the tutorial will set and rename your datum planes. To utilize datum planes within a drawing, and with geometric tolerances, you must first set them. The Set Up >> Geom Tol option is used to set datums in Part mode. The Edit >> Properties option in Drawing mode will be used in this tutorial to set datums.

Figure 6-49   Datum Display toolbar

STEP 1:   **On the toolbar, turn ON the display of datums planes (Figure 6–49).**

STEP 2:   **On the work screen, pick datum plane FRONT.**

Select datum plane FRONT. If during the construction process the location of your datum plane FRONT does not match the location of FRONT as shown previously in Figure 6–43, adjust your datum selection to match this figure.

STEP 3:   **On the menu bar, select EDIT >> PROPERTIES.**

After selecting this datum plane, the Datum dialog box will appear (Figure 6–50).

STEP 4:   **Rename this datum plane A.**

As shown in Figure 6–50, in the Name textbox, rename FRONT to a value of A.

STEP 5:   **On the Datum dialog box, set the datum by selecting the datum plane symbol button (Figure 6–50).**

If your datum plane symbol does not appear as shown in the figure, you did not properly set the drawing setup file option *gtol_datums* to *STD_ISO* as required earlier in this tutorial.

STEP 6:   **Select OKAY on the Datum dialog box to accept the values.**

After selecting OKAY, notice on the work screen how the datum symbol is now displayed in the ISO format (Figure 6–51)

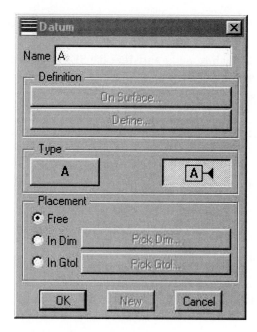

Figure 6-50   Datum dialog box

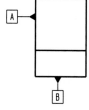

Figure 6-51   ISO datum
plane symbol

STEP 7: Drag datum Plane A to the location shown in Figure 6–51.

INSTRUCTIONAL NOTE   If your location of datum plane A does not match the location shown in Figure 6–51, you probably extruded the base protrusion of the part the wrong direction. This can be fixed in Part mode by redefining the direction of extrusion.

STEP 8: Use the EDIT >> PROPERTIES option to set datum plane TOP and rename it B (Figure 6–52).

STEP 9: Use the EDIT >> PROPERTIES option to set datum plane RIGHT and rename it C (Figure 6–52).

STEP 10: Select the VIEW >> SHOW AND ERASE option.

As shown in Figure 6–53, two datum plane B symbols are available in the drawing. Except for the need for clarity, only one datum plane symbol is required. You will use the Show/Erase dialog box to erase the second datum plane B symbol.

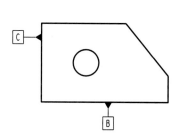

Figure 6-52   Setting datum planes

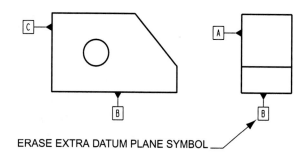

ERASE EXTRA DATUM PLANE SYMBOL

Figure 6-53   Erase the extra datum plane symbol

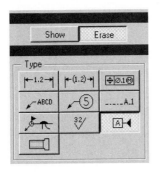

Figure 6-54 Datum plane item

STEP 11: On the Show/Erase dialog box, select the ERASE option, and then select the DATUM type (Figure 6–54).

STEP 12: On the Show/Erase dialog box, make sure that the SELECTED ITEMS Erase By item type is selected.

The Selected Items option will erase only items that are selected on the work screen.

STEP 13: On the work screen, select the extra datum plane symbol B (Figure 6–53); then select Done Select (Done Sel) from the Get Select menu.

STEP 14: Close the Show/Erase dialog box.

CREATING DIMENSIONS

This segment of the tutorial will show the parametric dimensions that were created when the model was constructed in Part mode. You will use the Show/Erase option to show all dimensions and centerlines.

STEP 1: Select the SHOW/ERASE option.

The Show/Erase option is used to show or not show items that were defined in Part and/or Assembly modes.

STEP 2: On the Show/Erase dialog box, select the SHOW option and deselect the DATUM option (Figure 6–55).

From the previous segment of this tutorial, the Erase option and the Datum item type should still be selected. If they are, select the Show option and deselect the Datum option.

Figure 6-55 The Show/Erase dialog box

STEP 3: On the Show/Erase dialog box, select the DIMENSION and AXIS item types (Figure 6–55).

The Dimension item type option will show parametric dimensions that were created in Part mode. The Axis item type will create centerlines.

STEP 4: On the Show/Erase dialog box, under the Options tab, check the ERASED and NEVER SHOWN options.

The Erased option will show items that have previously been erased, and the Never Shown option will show items that have never been shown.

STEP 5: On the Show/Erase dialog box, under the Preview tab, uncheck the WITH PREVIEW option (Figure 6–56).

Due to the limited number of dimensions in this tutorial, you will not preview your dimensions.

STEP 6: Select the SHOW ALL option, then select YES to confirm the show all.

The Show All option will show all available item types. In this step of the tutorial, all dimensions available from the referenced model will be shown.

STEP 7: Close the Show/Erase dialog box.

Your drawing should appear similar to Figure 6–57. Previously, you set the Drawing Setup File option *tol_display* equal to a value of Yes. This created the tolerance display shown on your drawing.

STEP 8: Use Drawing mode's MOVE and MOVE TEXT capabilities and the Pop-Up menu's SWITCH VIEW and FLIP ARROWS options to reposition the dimensions to match Figure 6–57.

Use the following options from the Detail menu to reposition the dimensions on the work screen.

• **Move** The Move option is available through the selection of a dimension's text with the left mouse button. Once selected, the dimension, including text and witness lines, can be dragged with your curser.

• **Move Text** The Move Text option is available through the selection of a dimension's text with the left mouse button. Once selected, grip points are available that will allow you to move only the text of a dimension.

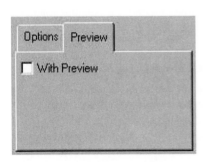

Figure 6-56 Showing without a preview

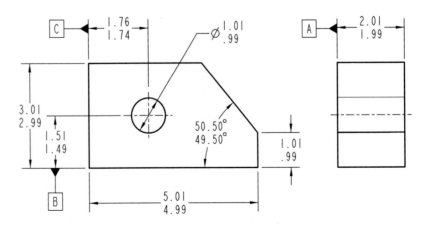

Figure 6-57 Drawing with dimensions and centerlines

- **Switch View** The Switch View option is used to switch a dimension from one view to another. It is available under the Pop-Up menu. Multiple dimensions can be selected by a combination of the Shift key and the left mouse button. Once dimensions are preselected, the Pop-Up menu is available through the right mouse button.

- **Flip Arrows** The Flip Arrows option is used to flip dimension arrowheads. It is available under the Pop-Up menu.

STEP 9: **Use Drawing mode's Move and Drag capabilities to create dimension witness line gaps.**

Where a dimension's witness line is linked to an entity, a visible gap (approximately 1/16-inch wide) is required. Drag the end of each witness line to create these gaps in each view.

STEP 10: Select the **CLEAN DIMENSIONS** icon.

The Clean Dimension option will consistently space dimensions on a drawing view.

STEP 11: **On the work screen, select the front and right-side views; then select Done Select (Done Sel) from the Get Select menu.**

Pro/ENGINEER's default values for the Clean Dimensions dialog box are 0.500 for the object Offset setting and 0.375 for the dimension Increment setting. You will keep these default settings.

STEP 12: **Select APPLY, then CLOSE the dialog box.**

After applying the clean dimension settings, Pro/ENGINEER will create snap lines that will evenly space linear dimension lines. Depending upon the complexity of the drawing, the applied settings might not create an ideal drawing.

STEP 13: **Use Dimension Modification tools to refine the placement of dimensions and datums.**

Use available tools such as Move View, Move, Flip Arrows, and Move Text to tweak the placement of dimensions. Your final dimensioning scheme should appear as shown in Figure 6–57.

STEP 14: **Save your drawing.**

SETTING GEOMETRIC TOLERANCES

Geometric tolerances are used to control the form and/or location of geometric features. Within Pro/ENGINEER, Geometric Tolerances can be incorporated into a model through Part, Assembly, or Drawing mode. This tutorial will first establish a Position tolerance ($\pm$.004 at MMC) for the 1-inch nominal diameter hole. Second, an Angular tolerance ($\pm$.005) will be provided for the 50-degree angled surface. Your final Geometric Tolerances should appear as shown in Figure 6–58.

STEP 1: **On the menu bar, select INSERT >> GEOMETRIC TOLERANCE.**

After selecting the Geometric Tolerance option, the Geometric Tolerance dialog box will be displayed. This is the same Geometric Tolerance dialog box that is utilized in Part mode. From this step forward, the steps for creating Geometric Tolerances are the same in Drawing mode as they are in Part mode.

STEP 2: **On the Geometric Tolerance dialog box, select the POSITION tolerance symbol (Figure 6–59).**

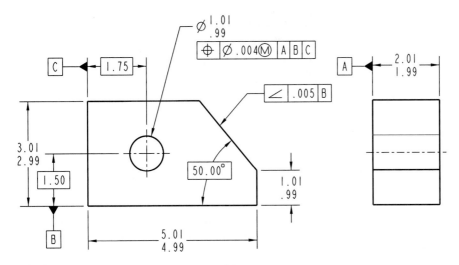

Figure 6-58 Drawing with geometric tolerances

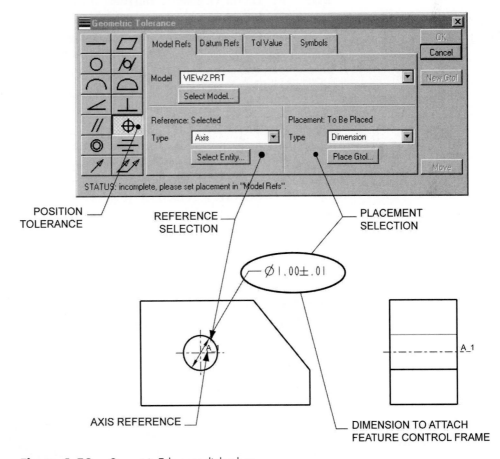

POSITION
TOLERANCE

REFERENCE
SELECTION

PLACEMENT
SELECTION

∅1.00±.01

AXIS REFERENCE

A_1

DIMENSION TO ATTACH
FEATURE CONTROL FRAME

Figure 6-59 Geometric Tolerance dialog box

STEP 3: On the toolbar, turn on the display of Axes (Figure 6–60).

This portion of the tutorial will create a Position tolerance for the 1-inch nominal size hole. A Position tolerance affects the axis of a hole. Due to this, in the next step of this tutorial you will be required to select the hole's axis.

Figure 6-60 Datum axis display

STEP 4:   On the Geometric Tolerance dialog box, change the Reference Type to AXIS (Figure 6–59).

STEP 5:   On the work screen select the axis at the center of the hole (Figure 6–59).

STEP 6:   On the Geometric Tolerance dialog box, set DIMENSION as the Placement Type (Figure 6–59).

STEP 7:   If necessary, select the PLACE GTOL option on the Geometric Tolerance dialog box.

STEP 8:   On the work screen select the 1-inch diameter dimension value (Figure 6–59).

When selecting the dimension value, you will have to select the dimension text. The nominal size of the hole is 1 inch. The dimension should be currently displayed with a Limit Tolerance format. This will be modified in a later step.

After performing this step, notice on the work screen how the Feature Control Frame is now visible.

STEP 9:   On the Geometric Tolerance dialog box, select the DATUM REFS tab.

STEP 10:   On the Geometric Tolerance dialog box, under the PRIMARY Datum Reference tab, set the Basic datum to A (Figure 6–61).

Notice how the Feature Control Frame updates dynamically.

STEP 11:   On the Geometric Tolerance dialog box, under the SECONDARY Datum Reference tab, set the Basic datum to B (Figure 6–62).

STEP 12:   On the Geometric Tolerance dialog box, under the TERTIARY Datum Reference tab, set the Basic datum to C.

STEP 13:   On the Geometric Tolerance dialog box, under the TOL VALUE tab, set the Overall Tolerance to a value of 0.004 and set the MATERIAL CONDITION to MMC (Figure 6–63).

STEP 14:   On the Geometric Tolerance dialog box, under the SYMBOLS tab, check the Diameter Symbol option (Figure 6–64).

Since a Position tolerance for a hole creates a cylindrical tolerance zone, it requires a diameter symbol with the tolerance value. Notice how the Feature Control Frame updates dynamically.

STEP 15:   Select OKAY on the Geometric Tolerance dialog box.

STEP 16:   If prompted, select YES to confirm the setting of Basic dimensions (if available).

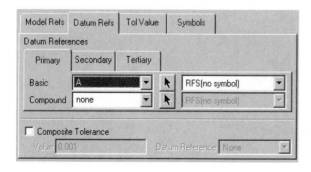

Figure 6-61   Primary datum selection

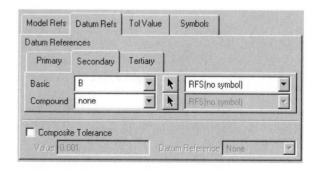

Figure 6-62   Secondary datum selection

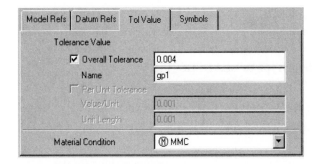

Figure 6-63   Tolerance value

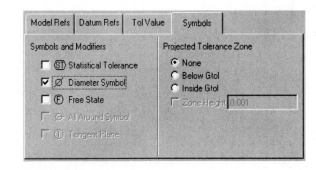

Figure 6-64   Diameter symbols selection

Your drawing should appear as shown in Figure 6–65. Notice how the value of each location dimension is enclosed in a box. This box represents a basic dimension. The confirmation required in this step sets the hole's location dimensions to Basic. If you do not get this message, use the Dimension Properties dialog box (Pop-Up menu >> Properties) to set each location dimension as basic.

STEP 17: **Select INSERT >> GEOMETRIC TOLERANCE.**

The remaining steps of this segment of the tutorial will create the Angular Geometric Tolerance.

STEP 18: **On the Geometric Tolerance dialog box, select the ANGULAR tolerance characteristic (Figure 6–66).**

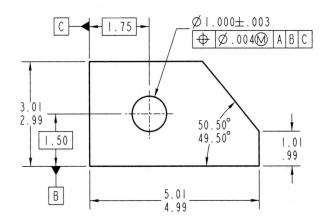

Figure 6-65   Position tolerance

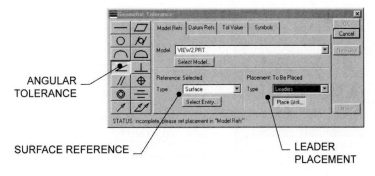

Figure 6-66   Geometric Tolerance dialog box

STEP 19: On the Geometric Tolerance dialog box, change the Reference type to SURFACE; then on the work screen select the edge of the angled surface.

STEP 20: Change the Placement Type to LEADERS (Figure 6–66).

The Feature Control Frame will be attached to the surface with a leader.

STEP 21: Select ON ENTITY >> ARROW HEAD on the Attachment Type menu.

STEP 22: On the work screen, pick the edge of the angled surface (only select once).

The selected location on the edge will be the attachment point for the leader. Only pick the edge once.

STEP 23: Select the DONE option on the Attachment Type menu.

STEP 24: On the work screen, select a location for the Feature Control Frame.

STEP 25: On the Geometric Tolerance dialog box, under the Datum Refs tab, set the PRIMARY datum reference to B (Figure 6–67).

An Angular Geometric tolerance requires one datum reference. In this example, datum plane B will be used as the datum reference.

STEP 26: On the Geometric Tolerance dialog box, under the Datum Refs tab, set the SECONDARY and TERTIARY datum references to NONE (Figure 6–67).

STEP 27: On the Geometric Tolerance dialog box, under the Tol Value tab, set the Overall Tolerance to a value of 0.005 and set the Material Condition to RFS [no symbol] (Figure 6–68).

— DATUM B

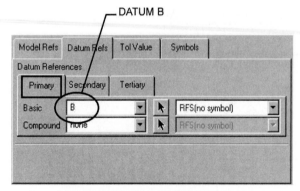

PRIMARY DATUM REFERENCE

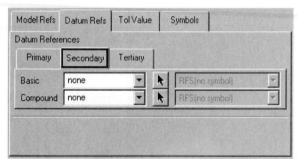

SECONDARY DATUM REFERENCE

Figure 6-67   Datum reference selection

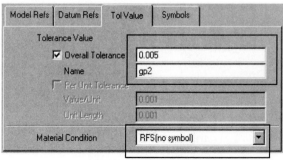

TOLERANCE VALUE

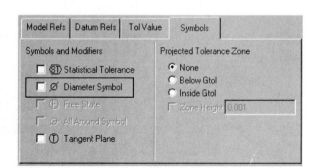

NO DIAMETER SYMBOL

Figure 6-68   Tolerance value and diameter symbol

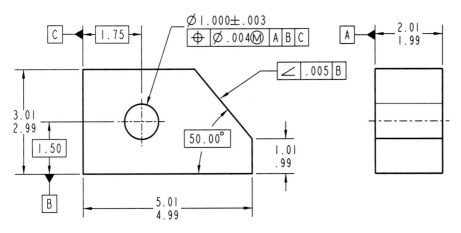

Figure 6-69   Geometric tolerances

STEP 28: On the Geometric Tolerance dialog box, under the Symbols tab, uncheck the DIAMETER SYMBOL button (Figure 6–68).

Angular Geometric tolerances do no create cylindrical tolerance zones. Hence, they do not require a diameter symbols.

STEP 29: Select OKAY on the Geometric Tolerance Dialog Box.

Due to the Angular Geometric tolerance, the dimension defining the size of the angle (50-degree nominal size) should be set as basic.

INSTRUCTIONAL NOTE   If you make a mistake with a geometric tolerance's feature control frame, use the Pop-Up menu's REDEFINE GTOL option to make necessary change.

STEP 30: Select the 50 angular dimension value with your left mouse button; then with your right mouse button, access the Pop-Up menu.

STEP 31: Select the PROPERTIES option on the Pop-Up menu to access the Dimension Properties dialog box.

STEP 32: On the Properties tab, select BASIC as the dimension display type.

STEP 33: Select OKAY to exit the dialog box.

Your drawing should appear as shown in Figure 6–69.

SETTING DIMENSIONAL TOLERANCES

This segment of the tutorial will create the dimensional tolerances shown in Figure 6–70. Previously in this exercise, you set the Drawing Setup File option *display_tol* to Yes. This option is used to display dimensions in a tolerance format.

STEP 1: On the work screen, select the hole's diameter dimension value; then access the Pop-Up menu with the right mouse button.

This portion of the tutorial will create a tolerance dimension defining the size of the hole.

STEP 2: Select the PROPERTIES option on the Pop-Up menu.

STEP 3: On the Dimension Properties dialog box, change the Tolerance Mode to +− SYMMETRIC (Figure 6–71).

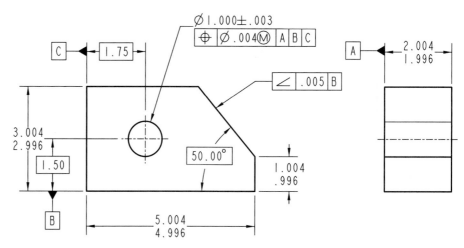

Figure 6-70   Dimensional tolerances

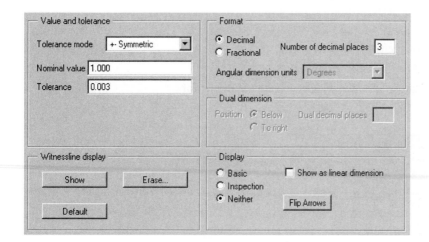

Figure 6-71   Modify Dimension dialog box

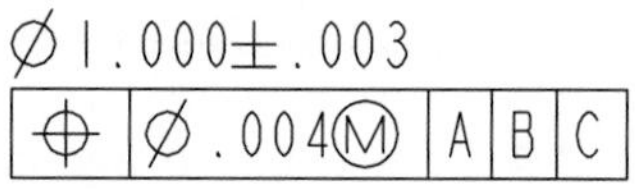

$\emptyset 1.000 \pm .003$

Figure 6-72   Hole dimension note

STEP   4:   On the Dimension Properties dialog box, change the Number of Digits to a value of 3 (Figure 6–71).

STEP   5:   On the dialog box, change the Tolerance value to 0.003 (Figure 6–71).

This will change the tolerance of the hole dimension to plus-minus 0.003 inches.

STEP   6:   Select OKAY to exit the Dimension Properties dialog box.

Your dimension note should appear as shown in Figure 6–72.

STEP   7:   Using a combination of your SHIFT key and left mouse button, pick the remaining four dimensions that are currently displayed with a Limit format (see Figure 6–73).

This portion of the tutorial will modify the remaining dimension to have a tolerance value of plus-minus 0.004 with a Limit format.

STEP   8:   Select EDIT >> PROPERTIES or select the Properties option on the Pop-Up menu.

STEP   9:   On the Dimension Properties dialog box, change the Number of Digits to a value of 3 (Figure 6–74).

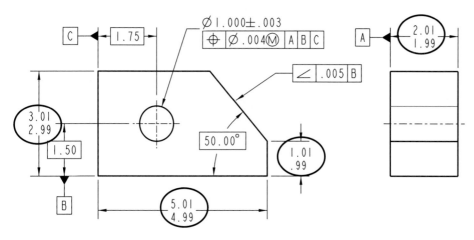

Figure 6-73 Dimensions to change format

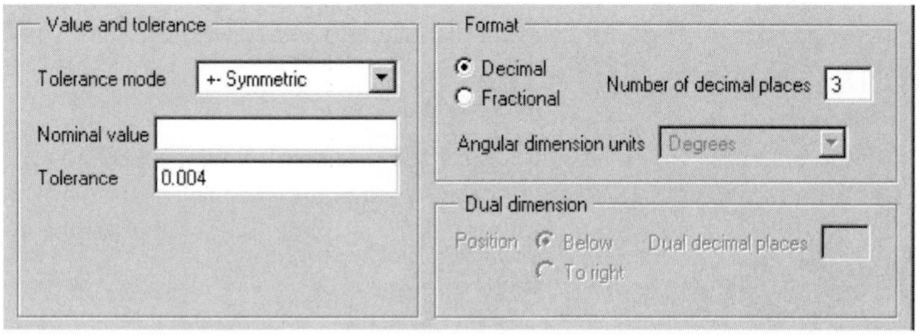

Figure 6-74 Modify Dimension dialog box

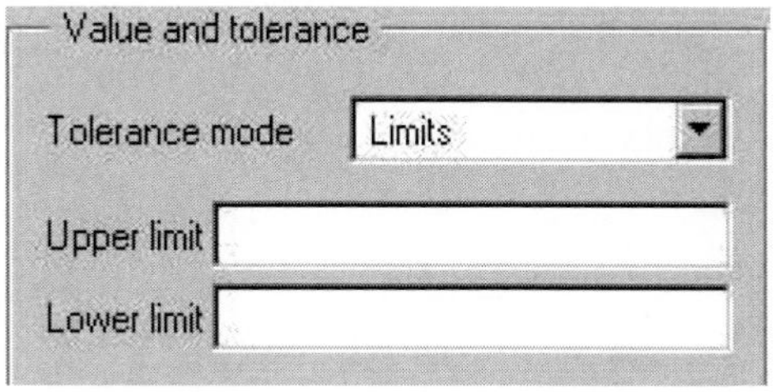

Figure 6-75 Limits tolerance format

STEP 10: On the Dimension Properties dialog box, change the Tolerance Mode to +− SYMMETRIC; then change the Tolerance value to 0.004 (Figure 6–74).

Changing the Tolerance Mode to +− Symmetric will allow you to change the Tolerance value for all four dimensions at once. In the next step, you will change the Tolerance Mode back to Limits.

STEP 11: Change the Tolerance Mode to LIMITS (Figure 6–75); then select OKAY on the Modify Dimension dialog box.

Your drawing should appear as shown in Figure 6–76.

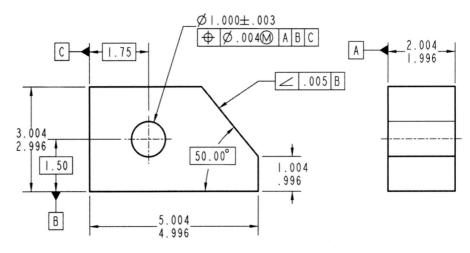

Figure 6-76   Final dimensioning and tolerance scheme

CREATING THE TITLE BLOCK

Use the Insert >> Note option to create the title block information shown in Figure 6–77. Move the scale note (*SCALE 0.5000*) to the title block. The Pop-Up menu's Modify Text Style (Mod Text Style) option can be used to modify the height and style of any text.

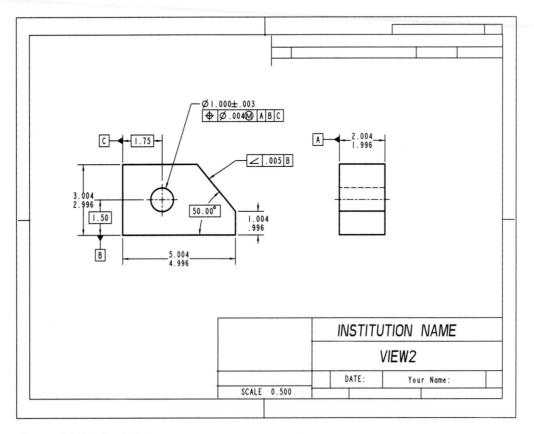

Figure 6-77   Finished drawing

PROBLEMS

1. Model the part shown in Figure 6–78. For this problem, meet the following requirements:
   - The dimensions shown in the figure meet design intent. During part modeling, incorporate these dimensions.
   - Create an engineering drawing with Front, Top, and Right-Side Views.
   - Use an A size sheet and format.
   - Fully dimension the engineering drawing using the part's parametric dimensions.

2. Model the part shown in Figure 6–79. For this problem, meet the following requirements:
   - The dimensions shown in the figure meet design intent. During part modeling, incorporate these dimensions.
   - Create an engineering drawing with Front, Top, and Right-Side Views.
   - Use an A size sheet and format.
   - Fully dimension the engineering drawing using the part's parametric dimensions.

3. Model the part shown in Figure 6–80. For this problem, meet the following requirements:
   - The dimensions shown in the figure meet design intent. During part modeling, incorporate these dimensions.
   - Create an engineering drawing with Front, Top, and Right-Side Views.
   - Use an A size sheet and format.
   - Fully dimension the engineering drawing using the part's parametric dimensions.
   - Apply geometric tolerance annotations and ISO datum plane symbols as shown in the figure.

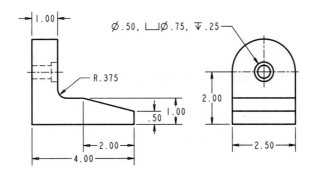

Figure 6-78   Problem one

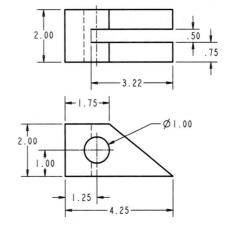

Figure 6-79   Problem two

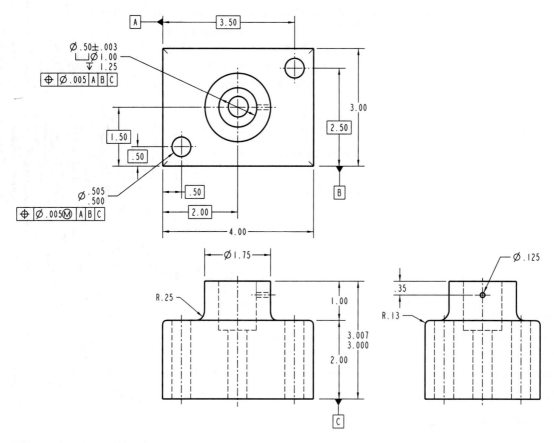

Figure 6-80   Problem three

QUESTIONS AND DISCUSSION

1. What is the default dimension text height of a Pro/ENGINEER drawing, and how can this text height be changed for an individual drawing? How can the default text height be changed permanently?

2. Describe the process used in Pro/ENGINEER's Drawing mode to add a border/title block overlay to a drawing. How are border/title block overlays created in Pro/ENGINEER?

3. Describe the difference between erasing a drawing view and deleting a drawing view.

4. Describe the difference between a General View and a Projection View.

5. Describe the difference between the Drawing mode views Half, Partial, and Broken.

6. How can a hidden line display be set permanently in a specific drawing view?

7. How can a tangent edge display be turned off permanently in a specific drawing view?

8. Describe the process used in Drawing mode to display parametric dimensions and centerlines.

9. Describe the process for adding symbols to a dimension note.

7

SECTIONS AND ADVANCED DRAWING VIEWS

Introduction

Pro/ENGINEER's Drawing mode provides a variety of options for creating orthographic and detailed drawings. This chapter will explore some of the advanced views available in this mode. Highlighted will be section views and auxiliary views. Upon finishing this chapter, you will be able to

- Create a full section view in drawing mode.
- Create a half section.
- Create an offset section.
- Create a broken out section.
- Create an aligned view.
- Create an auxiliary view.

DEFINITIONS

Auxiliary view   Any orthographic view that is not one of the six principle views.

Cutting plane line   A thick line used to show the cutting pattern of a section view.

Drawing setup file   A file used to establish environmental settings for a drawing. Examples of possible settings include text height and arrowhead size.

Offset section   A section view whose cutting plane is offset to include more features within the section.

Section lining   A pattern used on a section view to show where a model is cut.

SECTION VIEW FUNDAMENTALS

Section views are utilized within a working drawing to show details of a design that would be difficult to view using traditional orthographic projection. Figure 7–1 shows an example of a Full section view. A section view simulates what a model would look like if it were actually cut apart. On the Section View, **section lining** is used to show where the part is cut. Industry standards provide section lining patterns for a variety of materials. The default section lining used in Pro/ENGINEER represents iron (see Figure 7–1). A section lining's line style, weight, spacing, and angle can be changed with the Edit >> Properties option.

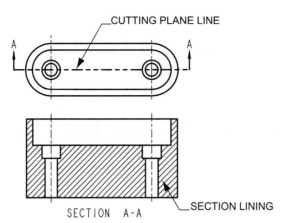

Figure 7-1   Full section view

Another important line type associated with a section view is the **cutting plane line.** Cutting plane lines typically lie in a view adjacent to the section view and represent the cutting path of the cross section (a Removed Section is an example of a view where the cutting plane line does not lie in a view adjacent to its section view). Arrowheads terminate the ends of a cutting plane line and point in the viewing direction of the section. The Drawing Setup File options *crossec_arrow_length* and *crossec_arrow_width* control the size of a cutting plane line's arrowheads.

Within Pro/ENGINEER, section views are created from defined Cross Sections. Pro/ENGINEER provides options for creating Cross Sections in Part, Assembly, and Drawing modes. Cross Sections created in Part and Assembly modes can be retrieved and used to create section views in Drawing mode. While creating a section view, Drawing mode also provides an option for creating a Cross Section.

SECTION VIEW TYPES

The most common section view used is the Full section. Other types of section views are available to serve a variety of documentation needs. The following is a description of the types of section views found under Pro/ENGINEER's Xsec Type menu.

FULL SECTION

The Full section view is the traditional type of section view used on most engineering drawings. As shown in Figure 7–2, A Full Section passes completely through a model. A Full section is available for General, Projection, and Auxiliary views.

HALF SECTION

The Half section view is similar to the Full section, except only half of the view is sectioned. As shown in Figure 7–2, for a symmetrical model, Half sections provide the advantage of a section on half the view, while also presenting the other half with traditional projection. A Half section is available for General, Projection, and Auxiliary views. It is not available with the Half, Broken, and Partial view types.

LOCAL

The Local Section option is used to create a Broken-Out section view. As shown in Figure 7–2, a Local section creates a section in a specific, user-defined area. Local sections are available for General, Projection, and Auxiliary views. It is not available with the Half and Broken view types.

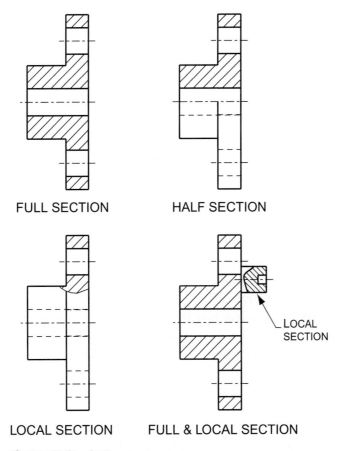

Figure 7-2 Section view types

FULL & LOCAL
_____

The Full & Local Section option is a section view with both a Full section and a Local section (Figure 7–2). The Full Section is placed first.

FULL SECTIONS

Full Section views cut completely through an object and show the entire model. Figures 7–1 and 7–2 show examples of Full Section views. Cross Sections used to construct a Full Section are created along a planar surface. Often, a datum plane is used as this surface. Perform the following steps to construct a Full Section view.

STEP 1: Select **VIEWS >> ADD VIEW.**

STEP 2: Select a View Type.

Full Section views can be created as a Projection, Auxiliary, General, Detailed, or Revolved view.

STEP 3: Select **FULL VIEW** as a view type.

Full View will create a view of an entire model.

STEP 4: Select **SECTION** from the View Type menu.

STEP 5: Select either **SCALE** or **NO SCALE** from the View Type menu.

For General views, the option is available for either specifying a view scale or for taking the default. When creating a section view that is projected

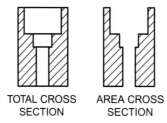

TOTAL CROSS
SECTION

AREA CROSS
SECTION

Figure 7-3   Total and area section views

from an existing view (i.e., Projection or Auxiliary), the scale of the parent view is used by default.

STEP 6:   **Select DONE to accept the view types.**

STEP 7:   **Select FULL as the section type.**

A Full section is a section that runs completely through a model (Figure 7–2).

STEP 8:   **Select TOTAL XSEC from the Cross Section menu.**

As shown in Figure 7–3, A Total Cross-section (Total Xsec) is a section view that includes geometry that borders the cross section and other geometry of the model. An Area Cross-section (Area Xsec) only shows the geometry that borders the cross section.

STEP 9:   **Select DONE to accept the Xsec Type values.**

STEP 10:   **On the work screen, select a location for the section view.**

STEP 11:   **Select CREATE from the Xsec Enter menu.**

> **INSTRUCTIONAL NOTE**   If an appropriate Cross Section already exists, use the Retrieve option to select the Cross Section.

Create will construct a Cross Section within the section view option, while Retrieve will select a Cross Section that has been previously created in Part or Assembly mode. The procedure for creating a Cross Section in Drawing mode is similar to the procedure for constructing a Cross Section in Part and Assembly modes. To construct a full section, a planar surface is needed. Datum planes are used often as this surface.

STEP 12:   **Select PLANAR >> DONE as the Cross Section creation method.**

Planar will create a Cross Section through a part at the location of a planar surface. While Planar is used to create a straight Cross Section, the Offset option will create a Cross Section that does not lie along a straight line.

STEP 13:   **Enter a name for the section view.**

In Pro/ENGINEER's textbox, enter a name for the section view. Within the area of mechanical drafting, sections are often named with an alphabetic character.

STEP 14:   **Select either a planar surface or a datum plane.**

This plane will create the cross section that will define the section view. The plane has to lie parallel to the section view location.

STEP 15:   **Select a view to locate the Cutting Plane Line (Figure 7–4).**

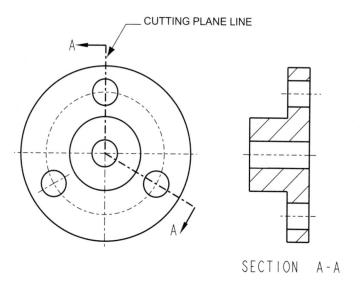

Figure 7-4 Cutting plane line

Pro/ENGINEER's message area is requesting a "view for arrows where the section is perpendicular." The Cutting Plane Line defining the Cross Section cut will be located in this view. If a Cutting Plane Line is not necessary, select the middle mouse button.

OFFSET SECTIONS

Most section views are situated along straight cutting planes. Often, features cannot be fully described through a straight cut. Drafting standards allow for an Offset Cutting Plane. Figure 7–5 shows an example of an Offset section view with its corresponding offset

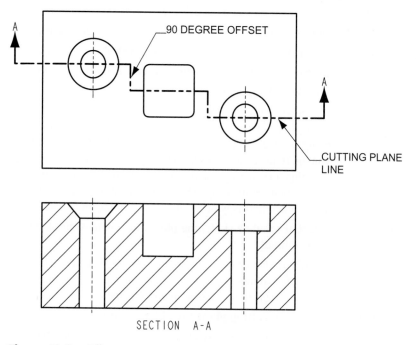

Figure 7-5 Offset section

cutting plane. With an **Offset section,** the cutting plane is offset by the use of 90-degree bends that allow the cutting plane to pass through features that require sectioning. While normal section views are created within Pro/ENGINEER using a planar surface or datum plane, the cutting plane line for an offset section is sketched. Perform the following steps to construct a projected offset section view.

STEP 1: Select VIEWS >> ADD VIEW >> PROJECTION.

Offset Section views can be created as a Projection, Auxiliary, or General view. This guide will demonstrate the creation of a section view as a Projection.

STEP 2: Select FULL VIEW as a view type.

STEP 3: Select SECTION from the View Type menu.

STEP 4: Select NO SCALE from the View Type menu (default selection).

For General views, the option is available for either specifying a view scale or for taking the default. When creating a section view that is projected from an existing view (e.g., Projection or Auxiliary), the scale of the parent view is used by default.

STEP 5: Select DONE from the View Type menu to accept the view type values.

STEP 6: Select a Cross Section Type, then select DONE.

Offset Section views can be created as a Full, Half, or Local cross section type. Additionally, a Total Cross-section (TotalXsec), Area Cross-Section (Area Xsec), Aligned Cross-Section (Align Xsec), or Total Align cross section type can be used.

STEP 7: On the work screen, select the location of the section view.

Use the mouse cursor to locate the section view. When creating an offset section as part of a General view, you are required to orient the view.

STEP 8: Select CREATE from the Cross-Section Enter menu.

Offset cross sections can be created in Drawing mode directly within the adding of a view, or they can be created in Part mode and retrieved during the placing of a section view. This guide will demonstrate the creation of an offset cross section view in Drawing mode.

STEP 9: Select OFFSET from the Cross-Section Create menu.

STEP 10: Select BOTH SIDES >> SINGLE >> DONE.

STEP 11: In Pro/ENGINEER's textbox, enter a name for the section view.

STEP 12: Switch to the model's window.

If necessary, use the Windows Taskbar or Pro/ENGINEER's Application Manager to switch to the part or assembly model from which the drawing is being created. When constructing an Offset section, for projection views, you are required to select a sketching plane on the actual model. When selecting a sketching plane from a projection view, ensure that the model from which the drawing is being produced is in the active window. You must switch to this window at this step in the view creation process.

STEP 13: Select or create a sketching plane, then orient the sketching environment.

Pro/ENGINEER creates offset sections by sketching a cutting plane line. This step requires you to select or create a planar surface that will be suitable for sketching the cutting plane line.

STEP 14: Sketch the cutting plane line (Figure 7–6).

Utilizing appropriate sketching tools, sketch the cutting plane line. As shown in the figure, when sketching the cutting plane line, ensure that you

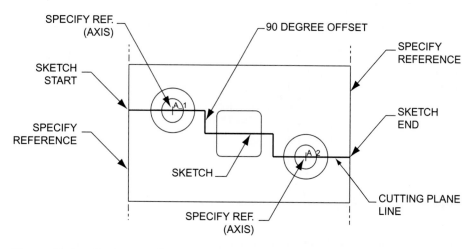

Figure 7-6 Sketching an offset section

differentiate between the lines that form the cut and the lines that form the offset.

MODELING POINT When sketching a cutting plane line with Intent Manager active, select the axis of each hole to include in the section as a reference. When sketching without Intent Manager, be sure to align the cutting plane line with the axis of each hole.

STEP 15: Select the Continue icon to exit the sketching environment.

STEP 16: On the drawing, select a view to display the cutting plane line.

STEP 17: Select OKAY to accept the direction of view or use the FLIP option to change the direction.

BROKEN OUT SECTIONS

Broken Out sections allow for the display of internal details of a model without the creation of a Full section. Figure 7–7 shows an example of a typical Broken Out section. Broken Out sections do not utilize a cutting plane line. Within Pro/ENGINEER, a Broken Out section can be created with a General, Projection, or Detail view. Perform the following steps to create a Broken Out section.

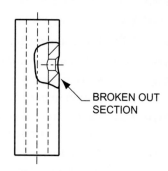

Figure 7-7 Broken out section

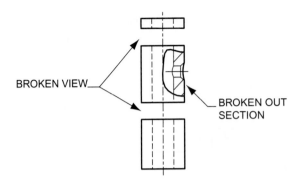

Figure 7-8   A broken out view versus a broken view

STEP 1: Select VIEWS >> ADD VIEW.

STEP 2: Select either GENERAL, PROJECTION, or DETAIL.

STEP 3: Select FULL VIEW >> SECTION.

Do not confuse the Broken View option with the creation of a Broken Out view. The Broken View option creates a view with break lines (Figure 7–8). Since this guide is demonstrating the creation of a Broken Out section on a full view, you should pick the Full View option.

STEP 4: If required, select either SCALE or NO SCALE.

STEP 5: Select DONE to accept view type values.

STEP 6: Select LOCAL >> TotalXsec >> DONE.

The Local option is the key selection for creating a Broken Out section. This option creates a section view within a sketched spline boundary. This boundary will be sketched in a later step.

STEP 7: On the work screen, select a location for the view, then (if required) enter a scale value.

From step 4, if you selected Scale, you must enter a scale value during this step.

STEP 8: Orient the Model (for a General View only).

For a General view, orient the model as required to properly display the Broken Out section.

STEP 9: Select ADD BREAKOUT >> SHOW OUTER on the View Boundary menu.

The Add Breakout option allows for the sketching of a spline boundary. The Show Outer option will show the remainder of the view outside of the sketched spline boundary.

STEP 10: Select CREATE on the Xsec Enter menu.

STEP 11: Select PLANAR >> SINGLE >> DONE.

STEP 12: Enter a name for the Broken Out section.

STEP 13: Select a plane for creating the Cross Section.

STEP 14: Select a view to place the cutting plane line or select the middle mouse button to not place.

STEP 15: On the work screen, select an entity approximately at the center of where the Broken Out view will be created (Figure 7–9).

This selection is required for the normal regeneration of the broken out portion of the sectioned view.

STEP 16: On the work screen, sketch a spline to create the boundary of the Broken Out section (Figure 7–9).

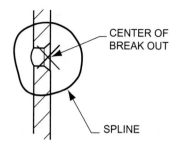

Figure 7-9   Creating the break out

Use the left mouse button to locate spline points and the middle mouse button to close the spline.

STEP 17:   **Select DONE on the View Boundary menu**

Multiple nonoverlapping splines may be sketched. Sketch as many as is necessary, then select the Done option. Your Broken Out section view will be created after selecting Done.

ALIGNED SECTION VIEWS

Engineering graphics is a language used to communicate design intent. Standards and conventions exist that govern the way designs are displayed on an engineering drawing. Within the realm of engineering graphics, designs are often displayed using multiview projection. Multiview projection does not always present the best display of a design. Figure 7–10 shows a design displayed using normal lines of projection. Normal lines of projection for a multiview drawing project at a 90-degree angle. With the drawing shown in the illustration, the part does not form a 90-degree angle. This presents a projection problem. Clarity for this design can be improved with the use of an Aligned view. As shown in Figure 7–11, the angled feature of the model can be aligned with normal lines of projection to create a drawing with more clarity.

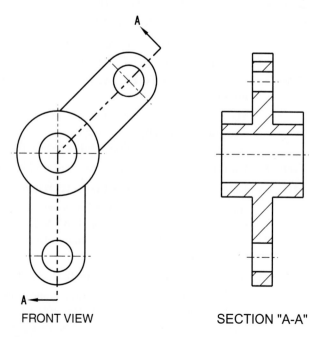

FRONT VIEW   SECTION "A-A"

Figure 7-10   Normal lines of projection

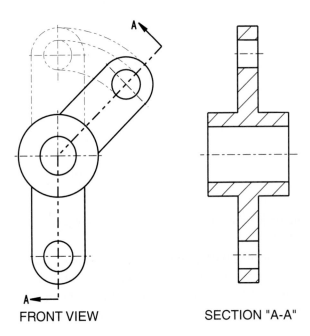

FRONT VIEW   SECTION "A-A"

Figure 7-11 Aligned view

Pro/ENGINEER's Drawing mode has the capability of producing an Aligned Section view. The following guide shows how to create the Aligned Section view shown in Figure 7–11.

INSTRUCTIONAL NOTE Cross Sections can be created in Part, Assembly, or Drawing mode. During the creation of a Section view, the Cross Section can be created during the View Add option. In this guide, the Cross Section was created in Part mode as an Offset Section. This Cross Section will be Retrieved during the view creation process.

Figure 7-12 Aligned xsec

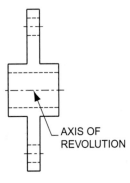

AXIS OF REVOLUTION

Figure 7-13 Axis of revolution

STEP 1: **Select VIEWS >> ADD VIEW.**

STEP 2: **Select PROJECTION >> FULL VIEW >> SECTION >> DONE.**

This example shows the creation of an Aligned Section view projected from an existing view. Aligned Section views can be created as a General view.

STEP 3: **Select FULL >> TOTAL ALIGN >> DONE on the Cross-Section Type menu.**

The Full option will create the section view completely through the model. The Total Aligned option will create the aligned view. Another option could be Align Xsec. This option would produce the view shown in Figure 7–12.

STEP 4: **On the work screen, select the location for the view.**

STEP 5: **Retrieve the section view that was created in Part mode.**

As previously mentioned, the Cross Section used in this guide was created in Part mode as an Offset Section. Pro/ENGINEER provides the capability for creating the identical Cross Section in Drawing mode through the use of the Create option.

STEP 6: **Select an Axis to revolve the feature about.**

This is the key step for the creation of an aligned view. In this example, select the axis shown in Figure 7–13. This axis serves as the rotation point for aligning the view.

STEP 7: **Select the view for the Cutting Plane Line.**

> In this example, the Cutting Plane line was placed in the Front view. If the cutting plane line is not desired, select the middle mouse button.

STEP 8: **Select OKAY to accept the default viewing direction.**

REVOLVED SECTIONS

Revolved Sections are used to show the cross section of a spoke, rail, or rib type feature. Additionally, they are used with features that are extruded, such as wide flange beams. Revolved sections are useful for representing the cross section of a feature without having to create a separate orthographic view. Revolved Sections are displayed by revolving the cross section 90 degrees. Figure 7–14 shows three different ways to locate the cross section in relation to its parent view. The following guide will demonstrate how to superimpose a Revolved section onto a view.

STEP 1: **Create or identify the view from which to obtain the Revolved Section.**

> A Revolved section can be created from a Projection, Auxiliary, or General view. Additionally, one can even be created from an existing Revolved view.

STEP 2: **Select VIEWS >> ADD VIEW >> REVOLVED.**

STEP 3: **Select FULL VIEW >> DONE.**

> The Revolved option allows for either a Full View or a Partial View only. A Revolved view by default has to be a section view.

STEP 4: **On the work screen, select the location for the Revolved view.**

> For a superimposed Revolved section view, select on the view used to create the Revolved section. The Move View option can be used to reposition the view.

STEP 5: **Select the view from which to create the Revolved Section.**

STEP 6: **RETRIEVE an existing Cross Section or CREATE a new one.**

> The Cross Section created or retrieved in this step will be used as the Revolved Section.

STEP 7: **Select a Symmetry Axis for the Revolved Section or select the middle mouse button to accept the default.**

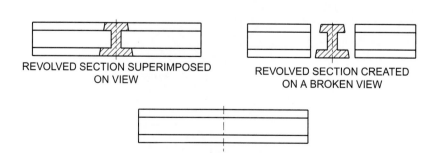

REVOLVED SECTION SUPERIMPOSED
ON VIEW

REVOLVED SECTION CREATED
ON A BROKEN VIEW

REVOLVED SECTION PLACED
OFF VIEW

Figure 7-14   Revolved sections

The Symmetry Axis is the location about which the Revolved Section view will be centered.

STEP 8: **Use the VIEWS >> MOVE VIEW option to refine the placement of the Revolved Section.**

AUXILIARY VIEWS

Within the language of engineering graphics, any object has six principle views. An **Auxiliary view** is any orthographic projected view that is not one of the six principle views. Auxiliary views are used frequently to show the true size of an inclined surface. Figure 7–15 shows an example of an auxiliary view and how it helps to better represent the inclined surface. This guide will demonstrate the creation of an Auxiliary view.

STEP 1: **Select VIEWS >> ADD VIEW.**

STEP 2: **Select AUXILIARY as a view type.**

Auxiliary is one of the five principle view types available in Drawing mode. An existing view is required before an Auxiliary view can be created.

STEP 3: **Select FULL VIEW >> NoXsec.**

An Auxiliary view can be created as a Half view or as a Partial view. Figure 7–16 shows an example of a Partial Auxiliary view. Additionally, an Auxiliary view can be created as a section.

STEP 4: **Select DONE from the View Type menu.**

STEP 5: **On the work screen, select a location for the Auxiliary view.**

STEP 6: **On the work screen, select an edge or an axis to project the Auxiliary view from (Figure 7–17).**

STEP 7: **Use the VIEWS >> MOVE VIEW option to reposition the Auxiliary view.**

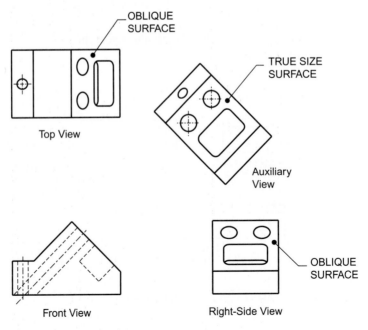

Figure 7-15   Auxiliary view

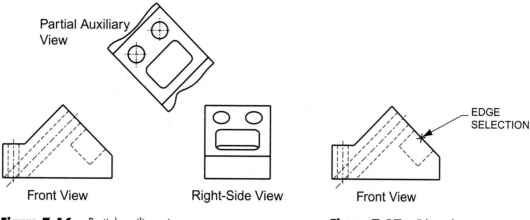

Figure 7-16   Partial auxiliary view   **Figure 7-17**   Edge selection

SUMMARY

Pro/ENGINEER is an integrated engineering design tool that allows for a full range of applications to include modeling, assembly, manufacturing, and analysis. Due to these tools, Pro/ENGINEER is a design package, not a drafting application. Despite this, there is still a need to document a design. Pro/ENGINEER's Drawing mode provides multiple tools for creating a high-quality engineering drawing. Included in these tools are the capability to create a variety of view types, such as section, detail, and auxiliary. Since Pro/ENGINEER is a fully associative computer-aided design application, models created in Part and Assembly modes can be used to create views in Drawing mode. Additionally, parametric dimensions that define a design can be used within Drawing mode to document the design.

ADVANCED DRAWING TUTORIAL

This tutorial exercise will provide instruction on how to create the drawing shown in Figure 7–18. The first step in this tutorial will require you to model the part from which the drawing will be created.

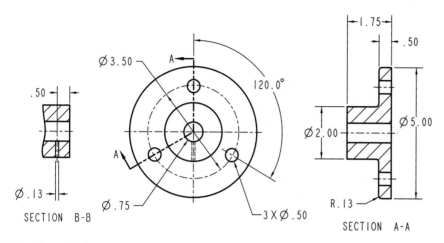

Figure 7-18   Completed views

Within this tutorial, the following topics will be covered:

- Starting a drawing.
- Establishing drawing setup values.
- Creating a general view.
- Creating an aligned section view.
- Creating a partial broken out section view.
- Annotating a drawing.

CREATING THE PART

Using Part mode, model the part shown in the drawing in Figure 7–18. Name the drawing file *section1*. Start the modeling process by creating Pro/ENGINEER's default datum planes. When defining features, the dimensioning scheme shown in the figure matches the design intent for the part. Incorporate this intent into your model. Create the base feature as a revolved Protrusion. Create the bolt circle hole pattern using the Radial Hole option and the Pattern command. The cross section for the drawing will be created in Drawing mode.

STARTING A DRAWING

This section of the tutorial will create the object file for the drawing to be completed. Do not start this section until you have completed the model portrayed in Figure 7–18.

STEP 1:   Start Pro/ENGINEER.

   If Pro/ENGINEER is not open, start the application.

STEP 2:   Set an appropriate Working Directory

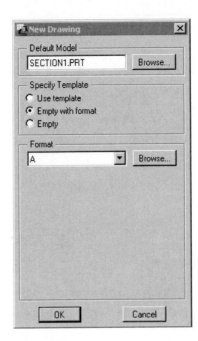

Figure 7-19   New Drawing dialog box

STEP 3:   **Select FILE >> NEW.**

STEP 4:   **In the New dialog box, select DRAWING mode; then enter *SECTION1* as the name of the drawing file.**

STEP 5:   **Select OKAY on the New dialog box.**

After selecting OKAY on the dialog box, Pro/ENGINEER will reveal the New Drawing dialog box. The New Drawing dialog box is used to select a model to create the drawing from and to set a sheet size and format.

STEP 6:   **Under the Default Model textbox, select the BROWSE option and locate the *SECTION1* part (Figure 7–19).**

Use the Browse option to locate the *section1* part created in the first section of this tutorial. The part will serve as the Default Model for the creation of drawing views.

STEP 7:   **Select the EMPTY WITH FORMAT option under the Specify Template section.**

You will use the Specify Sheet option to select an A size format.

STEP 8:   **Select the BROWSE option under the Format option on the New Drawing dialog box; then open a.frm as the Standard Sheet format (see Figure 7–19).**

The Browse option will allow you to browse the directory structure to locate an existing sheet format.

STEP 9:   **Select OKAY on the New Drawing dialog box.**

After selecting OKAY, Pro/ENGINEER will launch its Drawing mode.

ESTABLISHING DRAWING SETUP VALUES

This section of the tutorial will temporarily set Pro/ENGINEER's drawing settings. The Advanced >> Draw Setup option will be used to change the default settings for this specific drawing.

STEP 1: Select ADVANCED >> DRAW SETUP.

After selecting the Draw Setup option, Pro/ENGINEER will open the active drawing setup file with the Options dialog box.

STEP 2: **Change the Text and Arrowhead values for the current drawing.**

Within the Options dialog box, make the changes shown in the following table. The *radial_pattern_axis_circle* item is used to create a bolt circle centerline around the hole pattern.

Table 7-1   Values for drawing setup

Drawing Setup Item	New Value
drawing_text_height	0.125
text_width_factor	0.750
dim_leader_length	0.175
dim_text_gap	0.125
draw_arrow_length	0.125
draw_arrow_style	filled
draw_arrow_width	0.0416
radial_pattern_axis_circle	yes

STEP 3: **Apply the new options and close the dialog box.**

CREATING THE GENERAL VIEW

General views serve as the parent view for all views projected off of it. This section of the tutorial will create a General view as the Front view of the drawing (see Figures 7–18 and 7–20).

STEP 1: **Select the VIEWS menu option from the Drawing menu.**

STEP 2: **Select ADD VIEW >> GENERAL >> FULL VIEW >> NoXsec >> SCALE.**

STEP 3: **Select DONE on the View Type menu.**

STEP 4: **On the work screen, select the location for the Front view.**

The drawing under creation in this tutorial will consist of a Front view and a Full Section Right-Side view. The general view currently being defined will

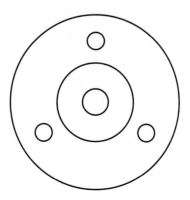

Figure 7-20   Front view

serve as the Front view. On the work screen, pick approximately where the Front view will be located. Once placed, the Views >> Move View option can be used to reposition the location of the view.

STEP 5: **In the textbox, enter .500 as the scale value for the view.**

STEP 6: **Use the Orientation dialog box to orient the view to match Figure 7–20.**

On the Orientation dialog box, use the appropriate references to set the view shown in Figure 7–20. Due to the nature of the part, you will have to select a datum plane as a reference. Pay careful attention that your hole locations match those shown in the illustration. If necessary, you can use the Views >> Modify View >> Reorient option to reorient the part.

STEP 7: **Save your drawing.**

Drawings are saved with an *.drw* file extension.

CREATING AN ALIGNED SECTION VIEW

This segment of the tutorial will create the Aligned Full Section view shown in Figure 7–21. The cutting plane for the section will match the path of the Cutting Plane Line shown. The Cross Section defining this section view will be created within Drawing mode as a part of this tutorial. Perform the following steps to create this view.

STEP 1: **Select VIEWS >> ADD VIEW.**

STEP 2: **Select PROJECTION >> FULL VIEW >> SECTION >> DONE.**

The Full View option will create a full projection view, while Section will provide options for creating a section view from the projected view.

STEP 3: **Select the FULL option on the Cross-Section Type menu.**

Do not confuse a Full section view with a Full View view type. The Full option from the Cross Section Type menu will create a section completely through the available view. This view could be a full, half, broken, or partial view.

STEP 4: **Select TOTAL ALIGN >> DONE on the Xsec Type menu**

The Total Align option allows for the creation of the Aligned Section view. The TotalXsec option would create a Full Section view, but the view would

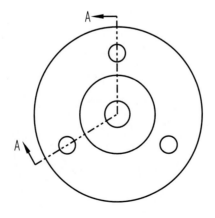
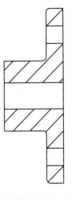

SECTION   A-A

Figure 7-21   Aligned section view

not be aligned. The Aligned Xsec option creates an aligned view also, but only the sectioning is displayed.

STEP 5: On the work screen, select a location for the Right-Side view.

STEP 6: Select the CREATE option on the Cross-Section Enter menu.

Pro/ENGINEER allows you to either create a new Cross Section or retrieve one that was previously created. In this tutorial you will create the Cross Section as an Offset.

STEP 7: Select OFFSET >> BOTH SIDES >> DONE.

Since the cutting plane for this section view does not follow a straight path, you will have to create an Offset section. An Offset section requires the sketching of the cutting path.

STEP 8: In Pro/ENGINEER's textbox, enter A as the name for the Cross Section.

STEP 9: Change to the window including the *section1* part (when using a Windows operating system, use the Taskbar to change windows).

Pro/ENGINEER requires you to sketch the cutting plane line within Part mode. If the part is currently open in a window, use the Windows Taskbar (or Pro/ENGINEER's Application Manager) to switch to the part's window. If the part is not open in a window prior to creating an offset section, Pro/ENGINEER will automatically open it for you.

MODELING POINT When a drawing is created from an existing model, the model (part or assembly) is placed in Pro/ENGINEER's session memory. Since the drawing is referencing the model, the model cannot be erased from session memory. Additionally, a model can be in memory and not be included in a window.

STEP 10: As shown in Figure 7–22, select the top of the part to use as the sketching plane; then select OKAY to accept the direction of viewing.

STEP 11: Select DEFAULT from the Sketch View menu to accept the default sketching environment orientation.

STEP 12: Specify the four references shown in Figure 7–23.

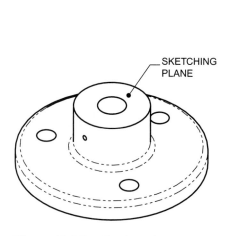

Figure 7-22 Sketching plane

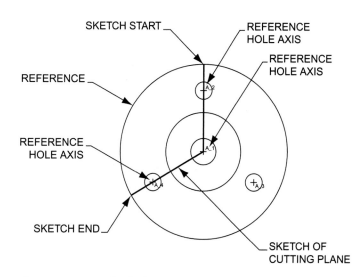

Figure 7-23 Sketched cutting plane

Notice in the figure how the three axes are specified as references. These axes will be used to define the path of the cutting plane. Dynamically rotating the model will make the selection of each axis easier (Control key and middle mouse button).

INSTRUCTIONAL POINT   If your part model was not opened before starting this segment of the tutorial, Pro/ENGINEER will open it automatically within its own subwindow. You will have to use the menu bar's Sketch >> References option to define the necessary references.

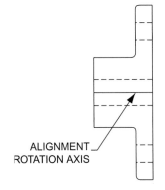

ALIGNMENT
ROTATION AXIS

Figure 7-24   Axis selection

STEP 13: Use the SKETCH >> LINE option to sketch the cutting path shown in Figure 7–23.

When sketching the path, make sure that each line is aligned with the references that were specified.

STEP 14: Select the Continue icon (or Sketch >> Done) to end the cutting plane's definition.

STEP 15: Switch to the Window including the drawing.

STEP 16: As shown in Figure 7–24, select the center axis to define the alignment rotation axis.

This selection will determine the pivot axis for the alignment.

STEP 17: On the work screen, select the Front view as the view to display the Cutting Plane Line.

In the message area, Pro/ENGINEER is asking you to "pick a view for arrows where the section is perp. MIDDLE button for none." Select the view to locate the Cutting Plane Line. If a Cutting Plane Line is not required, select the middle mouse button.

STEP 18: Select OKAY to accept the viewing direction.

CREATING A PARTIAL BROKEN OUT SECTION VIEW

This segment of the tutorial will create the Partial Broken Out view shown in Figure 7–25. Partial views are available with section and nonsectioned views. A Broken Out section view is created with the Local section type option and can be created as a Full, Broken, or Partial view.

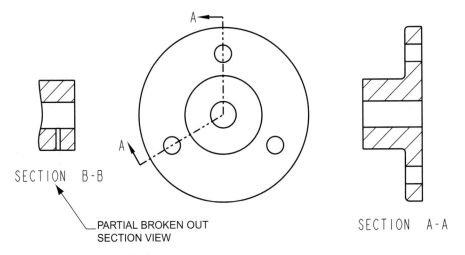

SECTION B-B

PARTIAL BROKEN OUT
SECTION VIEW

SECTION A-A

Figure 7-25   Partial broken out section

STEP 1: **Select VIEWS >> ADD VIEW.**

STEP 2: **Select PROJECTION >> PARTIAL VIEW from the View Type menu.**

The Partial View option requires you to sketch a spline to define the area that will be included in the view. The Local option that creates the Broken Out section also requires you to sketch a spline. This spline defines the area to include in the Broken Out section. When combining the Partial View and Local options, one sketched spline will serve for both functions.

STEP 3: **Select SECTION >> DONE.**

STEP 4: **Select LOCAL >> DONE on the Cross-Section Type menu.**

The Local option is used to create a Broken Out section view.

STEP 5: **On the work screen, select the location for the view.**

As shown in Figure 7–25, select a location that will project a Left-Side view.

STEP 6: **Select ADD BREAKOUT >> SHOW OUTER on the View Boundary menu.**

These options are selected by default. The Add Breakout option is used to allow for the sketching of a spline boundary, while the Show Outer option will show the sketched boundary in the view. At this time, do not select Done.

STEP 7: **Select the CREATE option on the Cross-Section Enter menu.**

You will create the Cross Section at this point in the tutorial.

STEP 8: **Select PLANAR >> SINGLE >> DONE on the Cross-Section Create menu.**

STEP 9: **In Pro/ENGINEER's textbox, enter B as the name for the Cross Section.**

STEP 10: **In the Front view of the drawing, select the datum plane that runs vertical through the drawing (see Figure 7–26).**

The datum as shown in the figure will serve as the planar surface that defines the cross section. Select a corresponding datum plane. If one does not currently exist on your model, use the MAKE DATUM option to create one on-the-fly.

STEP 11: **Select the middle mouse button.**

Pro/ENGINEER is requesting in the message area that you either select a view for the cutting plane line or select the "MIDDLE button for none." You will not use a cutting plane line for this section view.

STEP 12: **On the new Left-Side view, within the area to be sectioned, select a Center point for the outer boundary (Figure 7–27).**

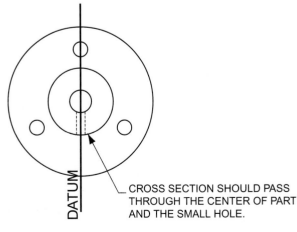

Figure 7-26   Cross section definition

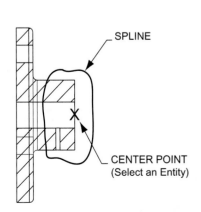

Figure 7-27   Sketching the boundary

On the work screen, in the Left-Side view, select a point on an entity that will reside in the partial section view. Pro/ENGINEER uses this selection to regenerate the section.

Note: If you have trouble selecting a point, dynamically zoom in on the view.

STEP 13:   **As shown in Figure 7–27, sketch a spline that will define the Partial View and the Local section.**

The left mouse button is used to select points on the spline, while the middle mouse button is used to terminate and close the spline.

STEP 14:   **On the View Boundary menu, select DONE.**

CENTERLINES AND DIMENSIONS

Within this segment of the tutorial, you will first show the centerlines shown in Figure 7–28. Second, you will display the parametric dimensions that define the part. Finally, you will set specific display modes for each view.

STEP   1:   **On the menu bar, select VIEW >> SHOW AND ERASE.**

STEP   2:   **On the Show/Erase dialog box, select the SHOW option and the AXIS Item Type (Figure 7–29).**

STEP   3:   **On the Show/Erase dialog box, select the SHOW ALL option; then select YES to confirm the selection.**

The Show All option will show all of a selected item type. In this example, all axes from the part will be projected as centerlines. Make sure that the Never Shown option is selected.

Note: If the With Preview option is selected, accept all of the centerlines shown on the work screen.

STEP   4:   **Use the Show/Erase dialog box to show all available dimensions.**

STEP   5:   **Use Drawing mode's MOVE and MOVE TEXT capabilities and the Pop-Up menu's SWITCH VIEW and FLIP ARROWS options to reposition the dimensions to match Figure 7–30.**

The Pop-Up menu is available by first selecting the text of a dimension with your left mouse button, followed by reselecting the same dimension text

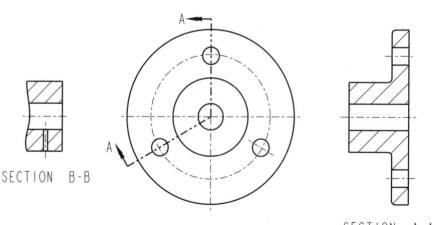

Figure 7-28   Centerlines

Figure 7-29   Show/Erase dialog box

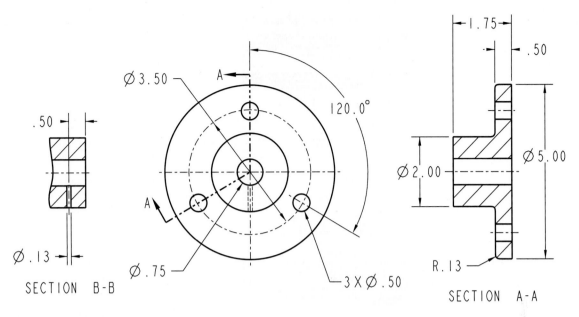

Figure 7–30   Drawing dimensions

with your right mouse button (*Note:* A slight delay in releasing the right mouse button is necessary). A variety of modification and manipulation tools are available through the revealed Pop-Up menu.

   Use the following options to reposition your dimensions on the work screen:

- **Move** The Move option is available through the selection of a dimension's text with the left mouse button. Once selected, the dimension, including text and witness lines, can be dragged with your curser.

- **Move Text** The Move Text option is available through the selection of a dimension's text with the left mouse button. Once selected, grip points are available that allow you to move only the text of a dimension.

- **Switch View** The Switch View option is used to switch a dimension from one view to another. It is available under the Pop-Up menu. Multiple dimensions can be selected by a combination of the Shift key and the left mouse button. Once dimensions are preselected, the Pop-Up menu is available through the right mouse button.

- **Flip Arrows** The Flip Arrows option is used to flip dimension arrowheads. It is available under the Pop-Up menu.

STEP 6: **Use the Show/Erase dialog box's ERASE option to hide dimensions not shown in Figure 7–30.**

STEP 7: ⬚ **Select the Clean Dimensions icon to set dimension spacing.**

The Clean Dimensions dialog box is used to consistently space dimensions.

STEP 8: **On the work screen, select the two section views; then select Done Select (Done Sel) on the Get Select menu.**

STEP 9: **On the Clean Dimensions dialog box, sets the options shown in Figure 7–31; then APPLY the settings.**

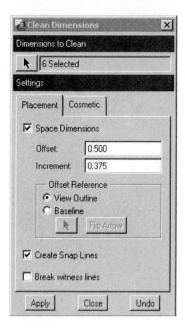

Figure 7-31   Clean Dimensions dialog box

STEP 10: **CLOSE the Clean Dimensions dialog box.**

STEP 11: **Select VIEWS >> DISP MODE >> VIEW DISP.**

The next exercise in this tutorial will have you set the display mode for each view. In this example, you will set the two section views with a No Hidden display. You will also set the Front view with a Hidden display and with a No Display Tangent display.

STEP 12: **On the work screen, select the two section views; then select Done Select (Done Sel) on the Get Select menu.**

You will set the two section views to not display hidden lines.

STEP 13: **Select NO HIDDEN >> DONE on the View Display menu.**

STEP 14: **Select VIEWS >> DISP MODE >> VIEW DISP.**

STEP 15: **On the work screen, select the front view; then select Done Select.**

STEP 16: **Select HIDDEN LINE >> NO DISP TAN >> DONE.**

You will set the front view to display hidden line and to not display tangent edges.

STEP 17: **With your left mouse button, pick the .50 diameter hole dimension then access the Pop-Up menu with your right mouse button.**

The next exercise in this tutorial will require you to modify the dimension text of the .500 diameter hole dimension to match Figure 7–32.

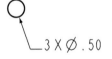

Figure 7-32   Dimension text modification

STEP 18: **Select the PROPERTIES option on the Pop-Up menu.**

STEP 19: **On the Dimension Properties dialog box, select the Dimension Text tab (Figure 7–33).**

STEP 20: **Modify the dimension parameters in the Dimension Text box (Figure 7–33) to add a 3X in front of the existing text.**

STEP 21: **Select OKAY to exit the Dimension Properties dialog box.**

STEP 22: **Save your drawing file.**

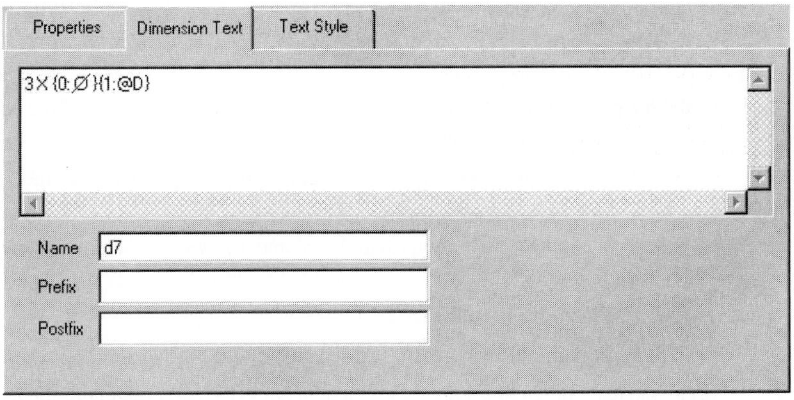

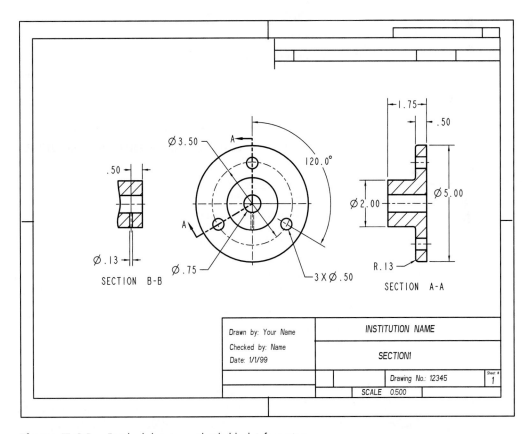

Figure 7-33 Dimension text tab

Figure 7-34 Finished drawing with title block information

TITLE BLOCK NOTES

In this segment of the tutorial, you will create the notes for the title block as shown in Figure 7–34. Use the Insert >> Note menu option to create the required text and the Pop-Up menu's Modify Text Style option to make text style and text height adjustments.

PROBLEMS

1. Model the part shown in Figure 7–35; then create a detailed drawing of the part. When completing this problem, meet the following requirements:

    •   The dimensions shown in the figure meet design intent. During part modeling, incorporate these dimensions.

    •   Create an engineering drawing with Front and Top views. The Front view should be a Full Section view.

    •   Use an A size sheet and format.

    •   Fully dimension the engineering drawing using the part's parametric dimensions.

2. Model the part shown in Figure 7–36; then create a detailed drawing of the part. When completing the problem, meet the following requirements:

    •   The dimensions shown in the figure meet design intent. During part modeling, incorporate these dimensions.

    •   Create an engineering drawing with Front and Top views. The Front view should be an Offset Full Section view.

    •   Use an A size sheet and format.

    •   Fully dimension the engineering drawing using the part's parametric dimensions.

3. Model the part shown in Figure 7–37; then create a detailed drawing of the part. When completing this problem, meet the following requirements:

    •   The dimensions shown in the figure meet design intent. During part modeling, incorporate these dimensions.

    •   When modeling the part, use the Radial Hole and Pattern commands to create the bolt-circle pattern.

    •   Create an engineering drawing with Front and Top views. The Front view should be a Half Section view.

    •   Use an A size sheet and format.

    •   Fully dimension the engineering drawing using the part's parametric dimensions.

4. Model the part shown in Figure 7–38; then create a detailed drawing of the part. When completing this problem, meet the following requirements:

    •   The dimensions shown in the figure meet design intent. During part modeling, incorporate these dimensions.

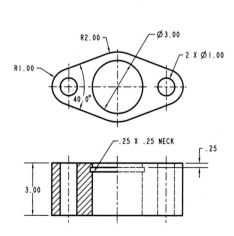

Figure 7-35   Problem one

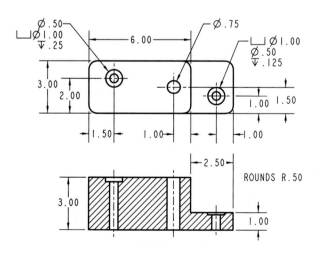

Figure 7-36   Problem two

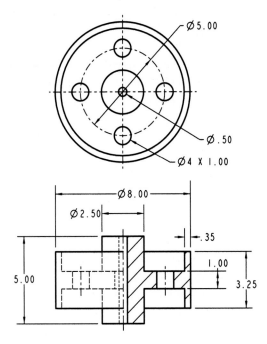

Figure 7-37   Problem three

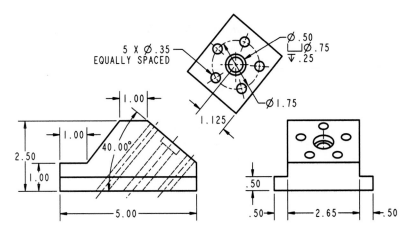

Figure 7-38   Problem four

- When modeling the part, use the Radial Hole and Pattern commands to create the bolt-circle pattern.
- Create an engineering drawing with Front, Top, Right-Side, and Auxiliary views. Create the Auxiliary view with the Of Surface view type option.
- Use an A size sheet and format.
- Fully dimension the engineering drawing using the part's parametric dimensions.

QUESTIONS AND DISCUSSION

1.   Describe possible uses of a section view.

2.   Describe different section view types available within Pro/ENGINEER.

3. What is the difference between a Total Cross Section and an Area Cross Section?

4. What is the purpose of a Cutting Plane Line?

5. In Pro/ENGINEER, how is the cutting plane created for an Offset Section?

6. In Pro/ENGINEER, what is the difference between a Broken view and a Broken Out view?

7. What is a Revolved section when used on an engineering drawing?

8. Describe uses of an Auxiliary view.

INDEX

WALTER SCHROEDER LIBRARY

3 9205 00065754 4

DISCARD